count down

BOOKS BY STEVE OLSON

Mapping Human History

Count Down

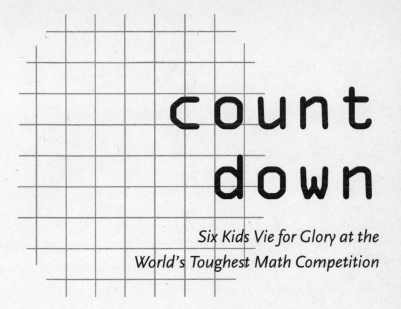

count
down

*Six Kids Vie for Glory at the
World's Toughest Math Competition*

STEVE OLSON

A MARINER BOOK
Houghton Mifflin Company
BOSTON   NEW YORK

FIRST MARINER BOOKS EDITION 2005

For information about permission to reproduce selections from
this book, write to Permissions, Houghton Mifflin Company,
215 Park Avenue South, New York, New York 10003.

Visit our Web site: www.houghtonmifflinbooks.com.

Library of Congress Cataloging-in-Publication Data

Olson, Steve, date.
Count down : six kids vie for glory at the world's toughest math competition / Steve
Olson
p. cm.
Includes index.
ISBN 0-618-25141-3
ISBN 0-618-56212-5 (pbk.)
1. International Mathematical Olympiad.   2. Mathematics—Competitions.
I. Title.
QA20.3.O47 2004
510′.79—dc22   2003056897

Printed in the United States of America

MP 10 9 8 7 6 5 4 3 2 1

MATHCOUNTS® is a registered trademark of the MATHCOUNTS Foundation,
1420 King St., Alexandria, VA 22314.

FOR DIANE OLSON

contents

count down

introduction

On July 4, 1974, a bus carrying eight U.S. high school students wound through the narrow medieval streets of Erfurt, East Germany. The students were all a bit nervous. In those days of heightened Cold War tensions, few Americans ventured beyond the Iron Curtain. Just that morning, after an all-night flight from New York City, the students had endured a brusque round of questioning by the East German border police. As they stepped from the bus in the center of Erfurt, beneath the spires of the cathedral where Martin Luther preached his first sermons, they felt both isolated and highly visible.

They were nervous for another reason. These high school juniors and seniors were the first team from the United States ever to compete in an International Mathematical Olympiad. In 1974 the Olympiad was already fifteen years old; the first one had been held in 1959 in Bucharest, Romania. But throughout the 1960s the United States had been reluctant to field an Olympiad team. The Olympiad is a competition for individuals in which gold, silver, and bronze medals are awarded. But unofficially the teams always have added their individual scores and compared themselves country against country. In this informal contest the Olympiad had been dominated by teams from the Soviet Union and eastern Europe. Even as more teams from western Europe began to compete — Finland in 1965 (finishing last), Great Britain, Sweden, Italy, and France (also finishing last) in 1967 — the

U.S. mathematics community had no desire to pit America's best high school students against the world's best. "A lot of people were dead set against it," says Murray Klamkin, a renowned mathematical problem writer who coached the U.S. team from 1975 to 1984. "They thought a U.S. team would be crushed by all those Communist countries."

In 1971 the mathematician Nura Turner, from the State University of New York at Albany, wrote an article that began to change people's minds. She pointed out that several state-level competitions, established mostly since the 1950s, had laid the groundwork for American participation at the international level. She admitted that a U.S. team might be humiliated in its initial attempts but argued that Americans were tough enough to bounce back. "We certainly must possess here in the USA the strength of character," she wrote, "to face defeat and the capability and courage to then plunge into systematic hard training to compete again with the desire to strive for a better showing."

In 1974 the major U.S. mathematical organizations finally agreed to send a team. Two years earlier the Mathematical Association of America had instituted a national exam designed to identify the best high school mathematicians in the country. In the spring of 1974 the association named the top eight finishers on the exam as the members of the U.S. Olympiad team.

Eric Lander, who is now one of the world's preeminent geneticists and the director of the Broad Institute of Harvard University and Massachusetts Institute of Technology, was a member of the team that first year. It was his senior year at Stuyvesant High School in Manhattan, and Lander was captain of the school's math team. "Math team was great," he says. "About thirty kids met each morning for an hour before school in a fifth-floor room of Stuyvesant High School, and the captain of the team was responsible for running the session. This was before you had databases full of math problems, so the captain of the math team, upon his ascension to office, came into possession of what we called 'the shopping bag.' It contained mimeographed sheets of problems and strips of problems and records of the city math contests for a long time. So the captain of the team would pull problems out of the bag and be responsible for leading the group."

When most people think about math competitions, they probably envision a roomful of kids struggling to perform complex calculations faster than the next person. But most of the problems in high-level competitions have very little to do with calculations. Solving these problems requires a sophisticated grasp of mathematical ideas, so that familiar concepts can be extended in new directions. The mathematical procedures everyone learns in school aren't enough. Becoming an excellent problem solver demands creativity, daring, and playfulness. A math competition is more like a game than a test — a game played with the mind.

The structure of an International Mathematical Olympiad reflects the nature of the problems. The size of the teams has changed over time. In the early years each team had eight members; since 1983 they have had six. But the format has stayed the same. On the first day of the competition all of the Olympians receive a sheet of paper containing three problems, and each competitor, working individually, has four and a half hours to make as much progress on the problems as he or she can. The next day they have the same amount of time to solve three additional problems.

But the competition doesn't begin when the competitors arrive in the Olympiad city, because the assembled team coaches first have to decide which problems will be on the exam. In Erfurt the teams had four days to tour the city and get to know one another.

"It was fascinating — the single team we most resembled and got along with were the Russians," says Lander. "So we hung out with the Russians a lot and got into all sorts of mischief.

We were in East Germany, and the Russians figured at that point that they owned East Germany, so they weren't going to get in trouble. I remember very well going up to the top of the dormitory at the school where we were staying, and the Americans and Russians throwing water balloons down on the street. The Russians might not do it back home, but they could do it in East Germany."

On July 8 the eighteen teams competing in the Sixteenth International Mathematical Olympiad gathered at a local university to take the exam. All the worries about the U.S. team's abilities had been for naught. Lander and his teammates finished second — just a few points behind the Soviet Union.

∇

This book is first and foremost the story of the Forty-second International Mathematical Olympiad, which took place in 2001 on the campus of George Mason University in Fairfax, Virginia, right outside Washington, D.C. The event has grown substantially since 1974. Nearly 500 kids from eighty-three countries competed in the Forty-second Olympiad, compared with about 125 in 1974 (and compared with the 150 or so who competed in 1981, the only previous Olympiad held in the United States). The Soviet team has splintered into teams from Russia, Latvia, Kazakhstan, and other former republics. Teams from South America and Africa — Argentina, Brazil, Colombia, Paraguay, Peru, Uruguay, Venezuela, Morocco, Tunisia, and South Africa — now compete. So do teams from East Asian countries such as Macau, Hong Kong, and the Philippines.

As one might expect, the competitors at the Forty-second Olympiad had their cultural differences, most notably the more than fifty languages that were spoken. But in general the Olympians were remarkably compatible. Most knew at least a little English, since English has become the language in which most of the world's higher-level mathematics is conducted. A soccer game immediately sprang up in the courtyard of the dormitory complex where they were staying and continued on and off for the duration of the event. All of the competitors could share CDs and hand-held video games, compare national qualifying exams, and lament the poor quality of the food offered in the college cafeteria.

Into this talkative, energetic, competitive mass of young mathematicians the U.S. team fit perfectly. Its members were fairly typical of those who had been on past U.S. teams. Five had just graduated from high school; one would begin his sophomore year that September. Three had spent at least part of their childhood in the San Francisco Bay area, two were from New Jersey, and one was from outside of Boston. Three participated in other team sports and were fairly athletic; the other three limited their athletic endeavors mostly to Ultimate Frisbee. All had been participating in math competitions at least since middle school.

If you had met the members of the U.S. team in a cafeteria or library or on the street, you wouldn't think there was anything special about them. They talked quickly and intensely among themselves, sometimes about math but usually about other subjects. They were rabidly interested in games of all sorts. They liked music, pizza, and movies.

But these kids *were* special. They were the products of one of the most intense selection processes undergone by any group of high school students. More than 15 million students attend public and private high schools in the United States, and nearly half a million take the first in a series of exams that culminates in the selection of the U.S. Olympiad team. The six individuals who emerge from that process are the best mathematical problem solvers of any American kids their age. Even someone who knew as much mathematics as they do would not have the benefit of the rigorous training the Olympians undergo.

What is it about the members of an Olympiad team that makes them such superb problem solvers? Some people would ascribe their talents simply to genius, saying that their accomplishments are so remarkable as to be beyond understanding. This use of the word "genius" as a label for the inexplicable has a long history. In classical Rome *genius* was the spirit associated with each individual from birth who shaped that person's character, conduct, and destiny. People sacrificed to their *genius* on their birthday, expecting that in return the guiding spirit would provide them with worldly success and intellectual power.

In the modern world the term often retains a hint of the supernatural. To call someone a genius is to imply that he or she is somehow distinct from normal human beings, with apparent access to experiences or thoughts that are denied to others. Genius from this perspective can seem to be, in the words of Harvard professor Marjorie Garber, "the post-Enlightenment equivalent of sainthood."

This way of thinking can skew even the most levelheaded analysis. In describing the achievements of the physicist Richard Feynman, Cornell University mathematician Mark Kac once made what has become a well-known distinction:

> There are two kinds of geniuses, the "ordinary" and the "magicians." An ordinary genius is a fellow that you and I would be just as good as, if we were only many times better. There is no mystery as to how his mind works. Once we understand what they have done, we feel certain that we, too, could have done it. It is different with the magicians. . . . The working of their minds is for all intents and purposes incomprehensible. Even after we understand what they have done, the process by which they have done it is completely dark.

Kac's distinction is beguiling, but it's really just a modern restatement of the Roman belief in spirits. Are the workings of some minds really incomprehensible? Or do great achievements rely on straightforward extensions of everyday thinking and imagining? Can profound advances in the arts and sciences be analyzed in such a way as to reveal their origins? Or are some realms of experience shut off from us forever, hidden behind the tantalizing veil of "genius"?

The varied meanings of the word complicate efforts to answer these questions. In modern parlance the term is often debased. People say that a politician is a genius at wooing voters. Newspapers label successful football coaches sports geniuses. Interior decorators, advertising writers, land developers, and country and western singers are all hailed as geniuses.

In middle and high schools, "genius" is usually a term of derision. The word is used to taunt someone who is good at math or a dedicated writer or simply more interested in schoolwork than the average student is. Even as adults, many people would feel uncomfortable being labeled a genius. The word seems an unwanted burden, a harbinger of unfulfilled expectations.

The kids on a U.S. Olympiad team would not consider themselves geniuses. They have become incredibly adept at solving immensely difficult mathematical problems. In that sense, they are *prodigies,* in that they have attained very high levels of performance at a young age. But they certainly are not geniuses in the sense that Homer, Archimedes, Shakespeare, Rembrandt, Newton, Mozart, or Einstein are so considered.

Nevertheless, the members of an Olympiad team do share the attributes of genius in one respect: they employ the same intellectual tools that history's great creators have. They use insight, talent, and creativity to produce original solutions to baffling problems. They exhibit the competitiveness, breadth, and sense of wonder that enable them to achieve at levels inconceivable to most people. By watching the Olympians solve mathematical problems, it's possible at least to glimpse the qualities that have produced humanity's greatest triumphs.

▽

Besides being about extraordinary achievements, this book is about mathematicians, a group that has received much attention in popular culture recently. Mathematicians have been the protagonists of hit movies (*Good Will Hunting, A Beautiful Mind*) and have figured prominently in well-received plays and novels (*Proof, Uncle Petros and Goldbach's Conjecture*). Princeton mathematician Andrew Wiles, who solved a famous mathematical problem called Fermat's last theorem in 1994, was even the inspiration for a musical in 2001 called *Fermat's Last Tango*.

This attention has been a mixed blessing. More than a few of these entertainments have made mathematicians out to be fools, nerds, or madmen. "Many recent works of mathematical fiction portray mathematicians as insane," says Alex Kasman, a mathematician at the College of Charleston in South Carolina, who maintains a Web site that reviews hundreds of fictional works involving mathematics. "Certainly there are mathematicians with mental illnesses, just as there are people of other professions with mental illnesses. But the high correlation of the two in fiction both supports and generates an unfair stereotype in the general population that there is some deep connection between the two. When I was watching *A Beautiful Mind,* and the character of John Nash was suffering terribly from his mental illness, I heard a woman behind me say, 'I'm glad I'm not a genius.'"

Other stereotypes plague works of fiction featuring mathematicians, says Kasman. Occasionally mathematicians are depicted as flamboyant and eccentric, like the "chaos theorist" played by Jeff Goldblum in the movie *Jurassic Park*. In other cases they are boring and repressed, like the husband (who also ends up deranged) in William Boyd's novel *Brazzaville Beach*. Rarely do moviegoers or novel readers encounter mathematicians with whom they might enjoy a conversation at a party. "I suppose no author wants to write about people who are ordinary," Kasman says. "So it's not surprising that very few fictional mathematicians are just ordinary people who like mathematics. But because most people do not personally know any mathematicians, they form their opinions of them based on these works of fiction. My experience, on the other hand, suggests that mathematicians are as normal as the people in any other profession."

One response to the stereotyping of mathematicians is to observe that scriptwriters and popular novelists stereotype *all* professions, even their own. But mathematicians have been absorbing abuse for a long time. In the Greek drama *The Birds*, written by Aristophanes in the fifth century B.C., a geometer named Meton arrives at a city founded by the Athenian Makedo and announces that he intends "to survey the plains of the air for you and to parcel them into lots." The populace denounces him as a "quack and imposter," beats him, and drives him from the city. In the novel *Emma*, published in 1815, Jane Austen asks whether a linguist, a grammarian, or "even a mathematician" could fail to appreciate the ardor of newfound love.

In American secondary schools, the stereotype of kids who are good at mathematics is somewhat different. They are seen as social misfits, physically uncoordinated, interested only in mathematics and other geeky subjects. Sometimes this stereotype turns up in television shows and movies as the badly dressed, awkward, computer-programming male who can't find a girlfriend.

The kids on an Olympiad team defy these brutally unfair stereotypes. Not all of them are interested in computers, science, or *Star Trek*. Some even claim to be not very good at mathematical calculations, at least compared with other Olympians. In fact, many of their traits initially seem antithetical to mathematics. They have deep insights into the problems they are solving. They are blindingly creative. They perceive the beauty in abstract mental constructs with an almost religious passion. And they are able to combine those traits in such a way that each trait builds on the others (though in this book I examine a different trait for each team member and each Olympiad problem).

None of the Olympians fits comfortably into the stereotype

of a mathematician. Each can be understood — and appreciated — only as an individual.

▽

Finally, this is a book about mathematics — about its complexities, its unreasonable effectiveness, its stark and breathtaking beauty. Many people believe that higher-level mathematics is conducted on a plane separate from normal thought, using concepts and logic that they could never hope to understand. To many mathematicians this belief seems misguided. They see mathematics as a smooth continuum from the numbers and shapes everyone learns in grade school to the frontiers of mathematical research. In many professions, acolytes need to make sudden leaps of achievement or skill, as when someone flies an airplane for the first time or teaches a class of boisterous students. Mathematics is not one of those professions.

A book about art has to include some reproductions of artwork, and a cookbook has to have recipes. By the same token, a book about problem solving should contain a few mathematical problems. For many people, the automatic reaction upon turning a page and seeing a geometric diagram or an equation will be "Oh no, not math!" That reaction is perfectly understandable. It arises from the boring mathematics classes most of us had to endure in school, the common belief that "I was never any good at math," and the widespread conviction that only the gifted few can hope to understand mathematics.

The six Olympiad problems in this book probably should be seen as extended examples rather than as core parts of the story. You don't have to understand the problems in detail to appreciate the skills that distinguish the Olympians. And readers who skip or skim over the problems will be in good company. When the English zoologist Sir Solly Zuckerman was asked once what he did when he came across mathematical formulas in scientific papers, he replied, "I hum them."

But anyone who can calculate a loan payment or a batting average is capable of understanding the problems described in this book. Olympiad problems are designed to involve only the mathematics that people learn in high school. They don't require a knowledge of subjects usually learned in college, such as calculus. Coming up with solutions to the problems is very challenging. The reason the Olympiad is generally considered the world's hardest mathematical competition is that high school students have relatively few tools with which to solve the problems, compared with older students who know more mathematics. Still, many of the solutions the Olympians devise are relatively easy to describe. For the three problems given on the first day of the Olympiad, the chapters of this book provide relatively complete solutions, with a few supporting details given in the appendix. For the second three problems, which are more complex, the chapters provide a general description of the solutions, with a somewhat more detailed treatment in the appendix. Working through one or more of the problems may take some time (though discussions of international relations, political gerrymandering, or the science of dieting are often more complicated), but the effort will be rewarded. As James Newman wrote in his classic anthology *The World of Mathematics,* "There are few gratifications comparable to that of keeping up with a demonstration and attaining the proof. It is for each man an act of creation, as if the discovery had never been made before."

Just as anyone can marvel at a great painting, a sublime piece of music, or a thunderous slam dunk without being a painter, composer, or basketball player, so anyone can appreciate the power and beauty of elegant mathematical problems and solutions. They are products of the human mind, as mysterious and inspiring as are all acts of creation.

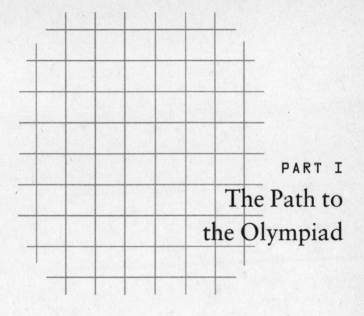

PART I

The Path to
the Olympiad

1 · inspiration

The fireworks spilled from the sky, right above the Washington Monument, as if they were being poured from a pitcher. The sound was tremendous — great booming explosions that rolled in waves across the Potomac River. On the upper deck of an ornate Mississippi River paddle wheeler named the *Cherry Blossom,* several hundred teenagers — in jeans and T-shirts mostly, bearing backpacks and cameras, obviously from many different countries — gazed upward at the spectacle. Several Chinese kids chattered merrily in Mandarin. Three or four Canadians tried to remember the lines to "America the Beautiful." But most of the observers were silent, staring slack-jawed at the fireworks display, the reds, greens, and yellows playing across their faces like half-forgotten ideas.

An hour before, the odds seemed slim that this show would get off the ground. As the *Cherry Blossom* paddled upriver from Alexandria, Virginia, a fierce storm swept across the Potomac, wrapping the ship in a thick blanket of swirling rain. The organizers of the cruise exchanged worried glances. Washington's Fourth of July fireworks show was supposed to be the inaugural event of the Forty-second International Mathematical Olympiad. Many of the competitors had endured long overnight flights specifically to be here for the fireworks. If the show was canceled, the Olympiad would be off to an inauspicious start.

But by dusk a sharp wind had scattered the clouds, unveiling a gibbous moon in the southwestern sky. Right above the city, however, the skies hadn't cleared. Throughout the show the fireworks burst first in and then below and then above a roiling cloud bank. The flashing waves of light gave the atmosphere a remarkable texture, a sort of weave, as if the clouds and winds were conspiring to reveal great and complex patterns.

Even before the sound of the final explosion had faded away, the young mathematicians on board the *Cherry Blossom* were filing belowdecks. The ship's ballroom was decorated to resemble a nineteenth-century saloon, but the Olympians paid no attention to the decor. They were intent on resuming what they had been doing for the entire trip upriver: playing games. The six members of the U.S. team were playing Association with the Bulgarians. The game begins with two players, one from each team, agreeing on a random word. One of the two then announces to all the rest of the players a word associated with the original word — but he has to be tricky, because if the clue is too obvious and his teammates fail to guess the word, the other team will almost certainly get it on the very next turn. In this game the word was "neck," and the U.S. team was going first. David Shin, a spiky-haired high school senior from West Orange, New Jersey, said "turtle." His U.S. teammates immediately answered "neck." In their somewhat fractured English, the Bulgarians objected — what possible connection could "turtle" have to "neck"? In Bulgarian the word for a turtleneck is "polo."

The next word was "shirt." Oaz Nir — handsome, trim, the son of Israeli parents who had immigrated to Louisiana — said "no." His teammates answered "yes." A kid on the Bulgarian team said "tea," and his teammates responded "coffee." Oaz immediately said "shoes," and his teammates answered "shirt." Wait, said the Bulgarians, how could you possibly get "shirt" out of "no" and "shoes"? The Americans replied, "No shirt, no shoes, no service."

Games like this were going on all around the room. The Koreans, cross-legged on the floor, were playing bridge and laughing over the hapless play of a teammate. Over by the stairs the Russians, joined occasionally by their former countrymen from Belarus and Latvia, were playing blackjack. The Irish were playing a game that they hadn't named yet.

Actually, "play" is too mild a word to describe what the people in this room were doing. They were competing. They leaned forward into the games, their eyes bright with calculation and strategy. Though many were still jet-lagged, they seemed full of energy, eager for a new game as soon as the previous game was won or lost. One got the sense that they could go on playing these games for many hours.

Sitting with the American team was an attractive, green-eyed, vivacious blond college student named Melanie Wood. When she was in high school, Melanie had been on the U.S. Olympiad team that finished third in Taiwan in 1998 and on the team that finished ninth in Bucharest in 1999. Now that she was in college and therefore no longer eligible for the Olympiad, she was instead serving as the U.S. team's guide. Each of the eighty-three teams had a guide, who stayed with the team throughout the Olympiad. The organizers tried to find a person who spoke the team's language; if that wasn't possible, they looked for someone who was affable and good at communicating without words. Over the next two days the guides would accompany their teams to the National Zoo and to a minor-league baseball game. After a rest day the competition would start on Sunday and conclude on Monday. During the week after the exam the guides and teams would visit the Smithsonian Institution, a nearby amusement park, and the science museum and aquarium in Baltimore. Meanwhile, teams of judges would be scoring the Olympians' papers. Nine days after the fireworks show, just up the Potomac at the Kennedy Center for the Performing Arts, the gold, silver, and bronze medals would be awarded.

The United States has sent a team to every Olympiad since

1974 (though the competition did not take place in 1980, when many international events, including the Olympic Games in Moscow, were disrupted because of the Soviet invasion of Afghanistan). Over that period 119 different students have represented the United States at the Olympiad. Here is the number of girls who have been on those teams: one. Melanie Wood is the only girl who has ever been a member of a U.S. Olympiad team.

Plenty of girls have qualified for the teams in other countries. Two members of the Finnish, the Australian, and the Trinidad and Tobago teams at the Forty-second Olympiad were girls. The teams from Denmark, Kuwait, and Thailand all had female members. Of the 473 competitors — representing probably the most talented group of young mathematicians in the world — 28 were girls. That wasn't a lot, but it was more than at many past competitions.

Everyone at an Olympiad has hypotheses about why women are underrepresented at the highest levels of mathematical achievement. Yet none of the explanations accounts for Melanie's experiences. She was born in 1981 in Indianapolis, Indiana. Her parents, Sherry Eggers and Archie Wood, had met at the public middle school where they both were teachers; Sherry taught Spanish and French; Archie was a math teacher. Archie had three daughters from a previous marriage, but they were all much older than Melanie, more like aunts than sisters.

When Melanie was six weeks old, her father died of cancer at the age of thirty-seven. Her mother, still teaching at the middle school, raised Melanie on her own. They spent long hours together doing errands. At home they read books and played games. By the time Melanie entered kindergarten, she was reading novels written for teenagers and knew how to multiply and divide. "My mom is a teacher, and she loved more than anything to teach," Melanie says. "She taught me everything she could."

Melanie attended good public schools in the eastern suburbs of Indianapolis and got good grades, but she did not specialize in math. "I can't recall ever thinking that she was superior in math compared with her other classes," says her mother, who later became principal of a middle school in Indianapolis. "She was bright in everything. I mean, she was reading difficult books in kindergarten, all those Baby-sitters Club books, the kinds of things that older kids read. And we used to play math games — actually I don't think of them as math games, just as games. You know, easy equations and things like that."

By the seventh grade, Melanie was in an accelerated math class. But she also excelled in many other areas; she was interested in drama, writing, and student government. (In high school she would be editor of the school newspaper, student government vice president, and a leader of the school's theatrical productions.) Outgoing and an easy conversationalist, she has always been extremely personable, with a wide range of friends and interests. Then one week in the spring of her seventh-grade year, she found her calling.

Every year thousands of middle school and junior high school students in the United States participate in a math competition known as Mathcounts, which was established in 1983 by the National Society of Professional Engineers, the National Council of Teachers of Mathematics, and the CNA insurance company. The idea was to give kids a way to hone their math skills and compete against other kids, just as spelling bees are designed to award achievement under pressure. In Mathcounts, four-person teams from individual schools compete at the chapter level, which usually consists of a city and the surrounding region. The highest scoring teams and individuals then compete at the state level, after which the four top scorers in each state form a team that goes to the national level. Though the competition is still largely unknown outside the group of people who care deeply about mathematics, it attracts intense interest among a highly dedicated subculture of students and their teachers.

In the seventh grade, Melanie wasn't part of that subculture;

she had never heard of Mathcounts. "I was asked the week of the competition — it was a Wednesday or a Thursday, I think — if I could go with the team to a math competition that weekend," she recalls. "The coach of the team had heard that there was this girl in the seventh grade who was in an advanced math class, and he figured that I would be okay at this." That Saturday she drove with the coach and the three other members of the team — eighth-graders whom she didn't really know — to a middle school in northern Indianapolis, where she competed against eighty or so other kids from the surrounding region. Melanie finished first in the competition. At the state competition a month later in Terre Haute, against the best seventh- and eighth-grade math students in Indiana, Melanie again finished first.

"My mom and I were just totally stunned by all this," she says. "I knew that I was pretty good at math, but I was pretty good at English and science and history — I was a good student. But I never would have guessed that I would win as a seventh-grader in the city, much less the state. That really flipped my world around in terms of making math something important in my life and changing my view of who I was and what I was good at."

At that point Melanie and three eighth-graders from other schools in Indiana had about a month to prepare for the national competition in Washington. In Mathcounts the coach of the team with the highest score in the state competition becomes the coach of the newly formed state team. Every year for more than a decade that coach had been Bob Fischer, a math teacher at Honey Creek Middle School in Terre Haute. Fischer had built championship teams year after year by getting hundreds of kids in his school enthusiastic about math, and from these groups a few kids inevitably emerged who went far in the competition. "I often tell my students that I'm not a gifted teacher," he says, "but I do have one gift. I can tell when you're working at your full potential."

Fischer had never heard of Melanie before she won the state competition. Now she was the only girl and the only seventh-grader on the Indiana state team. "It took only a couple of practices for me to realize that she was something special," says Fischer. "Her questioning and her depth, her understanding of not just the problems but the theory behind the problems, so that she could apply the concepts to much broader and more difficult problems — that was clear, even in the seventh grade. There's a book called *The Art of Problem Solving* — she got that at the state level as a gift. She sat down and read that book like a novel — she absorbed it that fast."

Before she started training with the state team, Melanie had never been around other kids who were extremely good at math. "The other students on the team thought and talked about math problems in the same way I did, and this was something I had never shared with any peers, or even teachers before. It was really exciting to be able to communicate about the problems I found so interesting to people who could understand my ideas. After one practice my mom said, 'You four were speaking another language.' Imagine what it would be like if you thought in a language other than English, and for the first time you found someone else who spoke that language!"

At the national competition Melanie finished fortieth of the 228 competitors. Of the thirty-four girls, she had the third-highest score. The next year, when she competed as an eighth-grader, the Indiana team was first in the nation, and Melanie finished tenth overall.

Melanie was twelve years old when she won the Indiana Mathcounts competition. A few months earlier she hadn't thought of herself as particularly distinguished in math. Where did her prodigious mathematical talents come from?

Several explanations come to mind. One is that she inherited her abilities from her math-teacher father — that her mathematical talents were somehow encoded in her genes. But purely genetic explanations for mathematical abilities immediately run into difficulties. Few well-known mathematicians are the children of mathematicians. Talent in mathematics seems to pop up out of nowhere and then fade within a generation or two. And Melanie's father was not a mathematician but a math teacher, though by all accounts a very good one.

Another possibility is that Melanie was somehow trying to compensate for the loss of her father through her interest in mathematics. She once told a reporter from a national science magazine, "[My father] is with me in the competitions, or even when I am just thinking about math. His spirit and his memory are there in my mind." Yet she concedes that his influence can easily be overstated. Her mother agrees. "Some people have said that she pursued this because her dad was a math teacher, and I've told her millions of stories about her dad and what a great person he was. But I don't think she feels the loss of him like I did. I think she was just glad that there was this good person who was her father. I don't think she was trying to do well because it would make her dad happy."

A third possibility — a suspicion often harbored about the parents of high-achieving children — is that her mother pushed Melanie to excel in mathematics. But this explanation, too, rings false. Talented kids who are pressured by their parents often end up turning away from both their talents and their parents, yet Melanie and her mother remain very close. Furthermore, her mother's success as an educator and her matter-of-fact approach to child rearing betray no traces of hyperambition.

"She's the only child I ever had," her mother says, "so I can't say I was ever surprised at anything she did. I assumed that she would be very bright. I know that sounds very stupid, but I can't think of any other way to say it. If I contributed anything to help her, it was teaching her that there isn't anything you can't do or be. And she still believes that. It's amazing the things she does. She's an amazing person. I'm very proud of her."

It isn't that these pop-psychology explanations are necessarily wrong. But they ignore the countless moments of chance and necessity that go into constructing a life. Melanie tells a story about something that happened when she was in kindergarten that obviously had a great effect on her. For as long as she can remember, she has been fascinated by a mathematical structure known as "Z mod 2Z." The term may sound forbidding, but it's really not that complicated. Z (the symbol comes from *Zahl*, the German word for "number") represents the whole numbers — 0, 1, 2, 3, and so on — along with the negative whole numbers, −1, −2, −3, and so on. When applied to Z, the operation "mod 2Z" sorts these numbers into two categories. One category corresponds to the odd numbers and the other to the even numbers. It is a mathematical analog of the ways we divide objects into contrasting dualities: light and dark, on and off, male and female, alive and dead.

"My earliest memory of this structure was when I was in kindergarten — and I got in trouble for it," Melanie says. "There were these little blue flash cards that had numbers on them from one to ten — I remember this so clearly — and I was playing with them and separating them into evens and odds. Maybe I'd heard the words 'even or odd,' but I hadn't really thought about it or understood it. And I was realizing things like when you added two odd numbers, no matter which two they were, you always got an even number, and when you added an even number to an odd number you got an odd number — things like that. And that's really the structure of Z mod 2Z. I got in trouble because I wasn't supposed to be playing with those flash cards. I'd already passed that level and was supposed to be playing with some other flash cards. I had probably just turned five, but I remember learning these facts about odd and even that were really just cool, just a very simple way of understanding the group structure of Z mod 2Z.

"Z mod 2Z is in many ways the simplest mathematical

structure. Yet you could study the mathematical properties of this structure for your entire life and you would never begin to understand it all. It's incredibly deep."

▽

Camilla Persson Benbow, dean of the Peabody College of Education and Human Development at Vanderbilt University, occupies the central suite of the Faye and Joe Wyatt Center, a magnificent Georgian office building that crowns a gentle slope on the south side of Nashville. The view from her window is of a perfect greensward of academic lawn, with the distant hills of central Tennessee just visible above the trees.

Benbow's academic career began almost three decades ago in Baltimore. As a senior at Johns Hopkins University, she had taken a job as a research assistant on a study of mathematically gifted junior high and high school kids. Each week she spent hours on the phone interviewing talented students and their parents. The director of the study was a psychology professor named Julian Stanley.

Stanley had become interested in mathematical precocity in the summer of 1968. Johns Hopkins had a summer program that brought middle school and high school students from the Baltimore area to the Hopkins campus to learn computer programming. One day an instructor in the program told Stanley about one of her students, an eighth-grader named Joseph Bates, who was teaching the instructors to program rather than the other way around. "I was somewhat hesitant and perhaps even reluctant at first to get involved [with a single student]," Stanley recalled at a 1992 conference held to honor his work. "There were too many other pressing duties. But I did, and my life and career were never to be the same."

Stanley first had to figure out what Bates could handle academically. Traditionally, this would have meant giving him an IQ test, but Stanley decided to take a different approach. Bates clearly was capable of doing college-level work, so Stanley decided to have him take the Scholastic Aptitude Test normally given to college-bound high school juniors and seniors. Neither Stanley nor Bates can remember Bates's exact scores, but they were "startlingly high," in Stanley's recollection.

At that point Bates was thirteen years old. One option was to move him up a grade, into high school. But his local high school would not let him enroll in the Advanced Placement courses he obviously needed. So Stanley again decided to innovate. After discussing various options with Bates's parents, he called the Hopkins admissions office. In the fall of 1969 Bates entered Johns Hopkins University as a college freshman.

Bates was far from the first youngster to enter a high-ranking college. Bright kids had been entering college early for many years, with mixed results. Many had found themselves isolated from peers, scorned by their classmates, and sometimes overwhelmed by the work. Bates had none of those difficulties. By the age of seventeen he had earned a master's degree in computer science from Hopkins, and he went on to become a distinguished research professor at Carnegie Mellon University in Pittsburgh. There he became involved in the Oz project, an effort to build computer software that combines characters, stories, and drama. More recently he cofounded Zoesis, a company in Newton, Massachusetts, that creates interactive cartoon characters for the World Wide Web with behaviors that mimic human emotions.

The experience with Bates piqued Stanley's curiosity. How many more kids were stuck in middle schools and high schools doing work that was obviously too easy for them? He wrote a grant proposal aimed at identifying and accelerating such kids, and in 1971 the newly created Spencer Foundation gave him a large award to carry out his plans. The next year Stanley sent letters to middle and junior high schools in the Baltimore area asking them to identify students who had previously scored in the top percentages on national standardized tests. About 450 seventh- and eighth-graders volunteered to take the SAT and other college-level tests. The next year Stanley contacted schools throughout the mid-Atlantic region; the year after that, letters went to schools throughout the country.

Stanley's interest in Joe Bates has led to one of the largest programs of educational acceleration in U.S. history. Today more than one hundred thousand seventh- and eighth-graders annually take the SAT or other high-level examinations. Few of the top scorers go directly to college — that route still tends to cause too many difficulties. But kids who do well on the tests qualify for summer programs, weekend courses at universities, independent study programs, travel abroad, and other enrichment opportunities. In the words of one researcher, "Psychology is often criticized for not being cumulative. Julian's work shows how it can be."

That initiative was still in its very early stages when Camilla Benbow began working with Stanley in 1976. Her first job was to help with a five-year follow-up study of middle-schoolers who had achieved especially high mathematics scores, which was the initial focus of Stanley's program. She found the work, and the kids, fascinating. "In December I talked with Julian and said that I would really be interested in staying on as a graduate student, and he loaded me up with books over the break — we had a really long break back then. I remember reading about Lewis Terman's fifty-year study of gifted students at Stanford, and I thought, 'Why can't I do this? I'm young enough.'"

Today Benbow codirects, with psychologist David Lubinski, the Study of Mathematically Precocious Youth, or SMPY, which is tracking more than five thousand of the students Stanley identified in the 1970s and 1980s. As in previous studies of academically advanced students, the subjects — now well into middle age — have had remarkably productive careers. More than a quarter have earned Ph.D.s, compared with about one percent of the U.S. adult population as a whole. They have published books, articles, and short stories. One has adapted Pink Floyd's *The Wall* into a multimedia rock opera. Others are video game developers, software engineers, violinists, corporate executives. According to Benbow, the group she has been studying has surpassed in achievement any other large group of children followed into adult life.

Yet the Study of Mathematically Precocious Youth has always had one notable idiosyncrasy. It has always been able to find a far greater number of mathematically talented boys than girls.

Stanley and his colleagues weren't expecting that. From the beginning they were focusing on students scoring in the top few percentages on national exams. At that level, they thought, the differences between boys and girls should be minimal.

The results of their first SAT test proved otherwise. Of the 396 seventh- and eighth-graders around Baltimore who took the test in March 1972, the 223 boys had an average score of 500 on the mathematics portion of the SAT. The 173 girls had an average score of 442. And the higher scores were even more skewed. In that first test, 43 boys scored above 600. Not a single girl did. Later tests identified higher-scoring girls, but the overall discrepancy remained. Among seventh- and eighth-graders who have taken the SAT, about twelve times as many boys as girls have scored above 700.

"SMPY was not formed to look for gender differences, but almost every woman who walks into the study is astounded by them," says Benbow. "And they're very specific differences. There're no differences in overall intelligence between girls and boys. It's a difference in relative strengths, It's the specific factor of math reasoning ability that seems to separate the sexes."

In 1980 Benbow published her first scientific paper, in *Science* magazine, with Julian Stanley as her coauthor. Titled "Sex Differences in Mathematical Ability: Fact or Artifact?" it observed that the differences between boys and girls could not be the result of differences in courses taken, because boys and girls in the seventh and eighth grades have taken essentially the same math courses. The paper acknowledged that different experiences outside the classroom might have "somewhat increased" the scores the boys received. But Benbow and Stanley clearly leaned toward a biological explanation for at least part of the test-score difference. "We favor the hypothesis that sex differences in achievement in and attitude toward mathematics result from superior male mathematical ability, which may in turn be related to greater male ability in spatial tasks," they wrote.

The article triggered what Benbow calls a "media field day." Newspaper and news magazine headlines shouted "Do Males Have a Math Gene?" (*Newsweek*) and "Are Boys Better at Math?" (*New York Times*). Other academics called the idea that boys were biologically favored in math "feeble," "fallacious," and "pseudoscientific." The controversy pushed Benbow even more firmly into the "biology matters" camp. In a long 1988 article in the journal *Behavioral and Brain Sciences,* she pointed toward several lines of evidence that suggest a biological advantage for boys. The same difference between boys and girls is found in other countries, even those in which the math culture is quite different. The differences appear at very early ages, perhaps even in preschool, though it depends on the specific ability being measured. Even girls who are equally interested in math score worse overall on tests than do boys. And odd biological correlates keep turning up: for example, among mathematically precocious children, a disproportionate percentage are left-handed, near-sighted, and prone to allergies, and left-handedness (though not the other two traits) is more common in boys than in girls.

Benbow generally has refrained from speculating about how a sex difference in mathematical abilities could have originated, but others have been less cautious. One prominent argument involves our evolutionary history. Several studies have shown that boys are better on average at imaging and manipulating shapes in their minds. Of course, say those who favor biological explanations of our traits; when human beings were evolving, the men hunted for meat while the women gathered plants. Men therefore evolved a keen spatial sense to find wild animals, to gauge the distance from their spears to their prey, and to throw their spears accurately.

This explanation has many flaws. Spears designed to be thrown were a relatively late invention in human societies; hunting was not necessarily a major source of food for early humans; and women also would have needed a strong spatial sense if they were gathering edible plants in the wild. Still, this explanation has a certain intuitive appeal. One study of boys in hunter-gatherer societies found that they throw sticks and rocks about three times as much as girls do, often at small animals.

Another popular explanation for the boy-girl discrepancy in math is that girls are more interested in people, presumably because they are practicing to be mothers, while boys are more interested in objects. For example, a somewhat controversial study conducted in the 1950s found that people were the subject of 80 percent of the stories told by two-year-old girls and only about 10 percent of the stories told by boys, who were much more likely to talk about objects like trains and cars. Another study found that baby girls pay much more attention to patterns that resemble facial expressions, whereas infant males are more interested in blinking lights, geometric patterns, and colored photographs of three-dimensional objects.

But any hypothesis about an inborn male advantage in math faces an almost insuperable obstacle. How can anyone tell if differences in observed behaviors have a biological or a social origin? Boys and girls at eighteen months of age show no difference in mathematical abilities, which even at that stage are remarkably advanced; human babies quickly develop a sense of "how much" and can make simple comparisons, say between three and two. But by eighteen months, parents and other adults have begun to treat boys and girls differently. Adults talk to boys and girls in different tones and using different words; these are generalizations, of course, but they apply in a surprising number of cases. Boys are given Erector sets and Legos, while girls are given tea sets and dolls. As soon as boys can walk they are encouraged to do sports, while rambunctious girls are urged to be more ladylike.

This socialization intensifies as soon as children go to school. Education researchers have found that teachers in U.S. elementary schools spend more time teaching math to boys than to girls. Boys more often play with scientific toys, participate in mathematical games, and read math books. And in many American households, when a child is having trouble with math homework, the mother's response is, "Go ask your father. I was never any good at math."

Despite these differences, sixth-grade boys and girls as a whole — though not those at the highest levels of achievement — average about the same on tests of mathematical ability. But then, at least in the United States, far more powerful social forces kick in. According to Melanie Wood's coach, Bob Fischer, "I think it starts in upper grade school and middle school. Girls hear that math is something boys do and girls don't do. Also, in middle school, girls start to feel that many boys won't like to be around them because they're too smart, so they pull back — I can see that happening."

Melanie has a somewhat different perspective. In middle school, kids begin to reflect on the social roles around them and how they will fit in, she says. That's when girls who enjoy math start to feel shut out. "There's no role in our society for girls who are good at math," she says. "I don't mean to imply that people who are good at math are nerds — many of them aren't. But it's easy to picture a male math nerd — there are movies about them, every TV show has a male nerd in there. There's no picture of what that would be for a girl. And because of the lack of that role in our society, a lot of girls don't see themselves that way."

Middle school is also when math becomes more competitive, and many girls dislike the added pressure. Girls have a tendency to prefer cooperative to competitive activities, studies of the classroom have shown. Boys who are skilled at math also begin to develop a certain macho air — "I can do problems that you can't." And they seem to be better at handling defeat, maybe because some of them have learned how to lose by playing sports. One study showed that when high school boys failed at an algebra problem, they increased their efforts, whereas high school girls tended to decrease their efforts.

The premier middle school competition, Mathcounts, also can be intimidating to girls. In the initial rounds, students solve problems individually and in their four-person teams. These rounds usually take place in small rooms without spectators or audiences. The final round is different. Called the Countdown Round, it is usually held in an auditorium before a large, enthusiastic crowd of parents, coaches, and reporters. Two at a time, the top-scoring students in the earlier rounds sit next to each other with their hands poised over buzzers. A problem such as the following is flashed on a screen:

What is the value of $2^3 \times 3 \times 5^3 \times 7$?

There is an easy way and there is a hard way to answer this question. The hard way is to multiply out $2^3 \times 3 \times 5^3 \times 7$, which equals $2 \times 2 \times 2 \times 3 \times 5 \times 5 \times 5 \times 7$. The easy way is to realize that the three 2s (2^3) can be multiplied by the three 5s (5^3) to yield three 10s. Therefore $2^3 \times 3 \times 5^3 \times 7$ is equal to $3 \times 7 \times 10 \times 10 \times 10$, which is equal to $21 \times 1,000$, or $21,000$. The typical national Countdown Round competitor would answer this question in less time than it takes most members of the audience to read it.

In the Countdown Round each question is allotted a maximum of forty-five seconds. The first person to hit the buzzer has three seconds to answer the question. If that person is wrong, the other competitor has the rest of the time to produce an answer. The first person to answer a specified number of problems correctly takes on a new challenger until a champion is crowned.

"The Countdown Round at Mathcounts was the most intense experience I've ever had," Melanie recalls. "In seventh grade, before I did my first Countdown Round, I didn't want to do it, but they told me that I had to. So I got up there, and after the very first question I had to run to the bathroom and throw up."

By the time U.S. students enter high school, the patterns are set. More boys than girls take advanced math classes such as calculus. Boys are more confident about their mathematical abilities, even though their grades in math are lower on average. Even many girls who are very good at math gradually lose interest and turn their attention to other fields.

These patterns have important consequences. Girls who learn to dislike math in high school are less likely to take math and science in college. As a result, they are less likely to become scientists or engineers or to go into other professions that require a technical background. Girls in high school and college then see relatively few women doing math and science and have to rely on the advice of men about whether to pursue an interest in those subjects. The cycle becomes self-reinforcing — and scientific studies that purport to find biological differences that make women less suited for such careers reinforce the status quo.

If policymakers and administrators decide that they want to encourage women to be more involved in mathematics, they have several options. They can give girls more incentives to study mathematics — for example, by providing college scholarships. They can give women preference in hiring decisions, so that female students have more role models and mentors. They even can set up separate classrooms for girls to teach math in a more nurturing way. But all of these options are susceptible to the criticism that they are a waste of time and money because "girls just aren't as smart in math."

Many people scoff at this stereotype and offer evidence to dispute it. In international comparisons of math achievement, girls in several other countries perform better on average than do U.S. boys. If mathematics instruction in the United States were better, they say, girls could do just as well as boys do now (though presumably the performance of the boys would also improve). Furthermore, social forces can quickly change the representation of men and women in an academic field. In 1970 just 5 percent of the students in medical schools were female — and plenty of men could explain why women were not cut out to be doctors. Today about half of the students in medical schools are women.

Even Camilla Benbow, who leans toward biological factors in explaining at least some of the disparity between boys and girls, says that opinions about the origins of those differences should not undercut reasonable efforts to eliminate them. "Whether the differences are biological or environmental in some ways doesn't matter, because our actions take place in the environment," she says. "We need to give girls equal opportunities to excel in math. Girls who have the talents and the interest should be encouraged and should see the excitement of math and science. And then they should be allowed to make the decisions that fit them best. I don't want to push people into careers that aren't satisfying to them. At the same time, I don't want to exclude people from careers because of some stereotype. I think girls should be given the opportunity to sample widely and make wise choices and not feel that certain things are out of their realm because of their gender."

▽

For many mathematically talented students, Mathcounts is just the first taste of high-level competition. In eighth grade many of these students take a test called the AMC 8 (AMC stands for American Mathematics Competitions, a program sponsored by the Mathematical Association of America). In high school they take the AMC 10 and AMC 12. A surprising number of students sign up for these competitions — about four hundred thousand annually take one of the AMC tests. From that group emerge the six individuals who will represent the United States at the International Mathematical Olympiad.

All the AMC tests have the same format. Students take them before or after school or during math class, usually with their math teachers serving as proctors. The high school tests consist of twenty-five multiple-choice problems, and students have an hour and a quarter to answer as many questions as they can. The problems start easy and get tough. For example, one question on a recent AMC 10 was

> What is the maximum number for the possible points of intersection of a circle and a triangle?
> (A) 2   (B) 3   (C) 4   (D) 5   (E) 6

This isn't a hard problem, but it helps to know some basic geometry. For any triangle, a circle can be drawn through the three points, or vertices, as shown by the triangles below with their circumcircles:

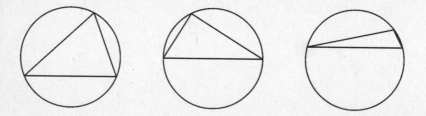

Now imagine each of these circles getting just a bit smaller while the triangle stays the same size. In that case the circle will cut through the triangle not at the vertices but at two points just inside each vertex. Thus the circle cuts through the triangle at six points, and the answer to the question is E.

Students who do extremely well on any of the AMC tests are invited to the next level of competition. In March of each year, copies of the American Invitational Mathematics Examination (AIME) are mailed to the teachers of about ten thousand qualifying students. The AIME consists of fifteen questions with whole-number answers, and students have three hours to solve them. As might be expected, the questions are appreciably harder than those on the AMC tests. Here's an example (the solution can be found in the appendix).

How many of the integers between 1 and 1,000, inclusive, can be expressed as the difference of the squares of two nonnegative integers?

At the next level of competition, about 250 of the top AIME finishers are invited to take the United States of America Mathematical Olympiad (USAMO), either at their own school or at a central location. The USAMO marks the beginning of a dramatic shift in the nature of the competitions. The test has the same format as an international Olympiad, with three problems to be solved in four and a half hours on each of the two days. Also, the problems no longer have multiple-choice or whole-number answers. Instead, they usually require that the competitors prove a mathematical result. Students can no longer excel by knowing a lot of mathematics and being pretty good with numbers. To do well on the USAMO, they need an additional set of mathematical skills.

A book called *The Art and Craft of Problem Solving* by former math Olympian Paul Zeitz illustrates this shift in perspective with the following example.

A monk starts to climb a mountain at 8:00 A.M. and reaches the summit at noon. He spends the rest of the day and that night on the summit. The next morning he leaves the summit at 8:00 A.M. and descends by the same route he used the day before, reaching the bottom at noon. Prove that there is a time between 8:00 A.M. and noon at which the monk was at exactly the same spot on the mountain on both days. Note that the monk can walk at different speeds, rest, or even go backward whenever he wants.

A student drilled in school mathematics would immediately try to change this problem into an equation. The monk could walk at speed v up the mountain. The time since he started could be labeled t. In that case, if he walked up the mountain at a constant speed, his distance up the mountain would be $v \times t$. But wait — the monk can walk at different speeds and can even go backward. How can that be factored into the equation?

The creative problem solver searches for alternate strategies. What are the essential elements of the problem? Can it be reframed to make it easier to solve? And then, whether through persistence, inspiration, or just plain luck, the flash of insight occurs. Pretend there is a second monk. As the first monk is coming down the mountain on the second day, the second monk travels up the mountain just as the first monk did the day before, with the same varying speeds and rests. The time when the two monks meet is the answer to the problem.

Students who do well on the USAMO are the superstars of high school mathematics. Each June the top twelve finishers travel to Washington and are feted at the State Department. In the ornate diplomatic reception room on the top floor of the building, overlooking the Mall and the Potomac River, government and industry officials make speeches and present special awards and scholarships. Right before dinner the students traditionally cross the street to the National Academy of Sciences to have their picture taken with the great bronze statue of Albert

Einstein that sits on the academy's front lawn. The pictures are posted on Web sites and hung on walls; among young mathematicians they are a badge of great distinction.

∇

In 1998, when she was a junior in high school, Melanie Wood earned the highest score in the nation on the USAMO. The next year she convinced most of the top finishers to wear tuxedos to the State Department dinner, and she wore a flowing white satin gown. In their picture with Albert Einstein, the USAMO winners look like kids at their high school prom, their faces lit up with excitement and triumph, about to celebrate the most festive day of their lives.

But for the country's best young mathematicians, the State Department dinner is an overture, not a finale. The real competition is just beginning.

2 · direction

On the morning of July 8, in the dorm rooms of George Mason University, many of the 473 competitors at the Forty-second Olympiad lay awake in bed, staring into the predawn darkness. They were as well prepared for the competition as they could be. They had been training for the Olympiad for months and years — for their whole lives, really. But now they knew that any lapse — a momentary loss of focus, a stupid mistake, a panic-stricken failure to perform — could mean the difference between success and failure, between acclaim and disgrace.

By 8:30 the competitors had eaten breakfast and were walking toward the Patriot Center, a ten-thousand-seat basketball arena on the George Mason campus. At the arena's south entrance they paused one last time to shake hands with their teammates and wish them good luck. Then they made their way to the 473 folding tables set up in huge rectangular arrays on the arena floor and in the concourses. Each of the tables was covered with a white tablecloth. Each bore a bottle of water, a granola bar, and a white envelope marked with the words DO NOT OPEN UNTIL INSTRUCTED TO DO SO. The competitors pulled out the chairs at their assigned tables and sat down. They jiggled their legs, chewed on their hair, gazed absently up toward the shadowy rafters. "This is going to be very loud," called a middle-aged woman in stretch pants and a purple sweater from the first row of seats, holding up an air horn. "You might want to cover your ears." Then she raised the air horn as high as she could and let loose a blast that echoed from the roof, reverberated through the concourses, and carried into the parking lots and woods outside.

The Olympians ripped open the envelopes and took out a single sheet of paper. On that sheet were three mathematical problems. The first involved a circle and a triangle, the second a short algebraic inequality, and the third a group of boys and girls taking a math test. An unmistakable sigh of relief issued from the competitors. These problems didn't look that hard. Maybe this test wasn't going to be the nightmare they'd heard it could be.

The Olympians were wrong about the test. Most professional mathematicians would not be able to solve these problems in the four and a half hours available to the Olympians.

▽

In January 1657 an unusual letter arrived on the desks of many of the leading mathematicians of Europe. Later referred to as "Two Mathematical Problems Posed as Insoluble to French, English, Dutch, and all Mathematicians of Europe by Monsieur de Fermat, Councillor of the King in the Parlement of Toulouse," the letter challenged the mathematicians to solve two specific problems: one involving possible ways of evenly dividing a cubed number; the other, possible ways of evenly dividing a squared number. In spirit the problems were not much different from the problem that would later become known as Fermat's last theorem.

The letter set off a ferocious tussle among Pierre de Fermat's correspondents. Some dismissed the problems as unimportant. Others sought to solve them and failed. After a lengthy and sometimes acrimonious correspondence, Fermat grudgingly acknowledged that he had received several solutions, yet he couldn't resist stoking the international rivalries of the day. "The French will say that the English satisfied the proposed problems," he wrote to a friend. "But let the English say in turn that the problems were worthy of being proposed to them and let them not disdain in the future to examine and investigate more closely the nature of integers."

In the seventeenth century, personal challenges were common in mathematics. At a time when many mathematicians were still amateurs (Fermat, for example, was a lawyer and jurist), they could make their reputations by solving problems that no one else had been able to solve. Many famous scholars of the time, including Isaac Newton and René Descartes, posed and worked on challenge problems. Many kept their procedures secret to maintain an advantage over their rivals.

In the eighteenth and nineteenth centuries, mathematicians gradually adopted the more decorous procedure of publishing their results in books, monographs, and journals. Yet the problem-solving tradition remained strong in some parts of Europe, especially as a way for young mathematicians to develop facility. Eastern Europe, in particular, remained a hotbed of mathematical competitions. In 1894, for example, a high school teacher in Hungary named Dániel Arany began publishing a student magazine called *KöMaL* (an acronym for "High School Mathematics Journal" in Hungarian). Each issue presented ten to twenty challenging problems that students were invited to solve, and Arany received hundreds of solutions each month in reply. Not coincidentally, many of the most accomplished mathematicians of the first half of the twentieth century came from eastern Europe — including many Jews fleeing Nazi Germany who came to the United States and made major contributions to the Allied war effort.

Since the end of World War II, problem-solving competitions as a form of mathematics education have spread around the world. But many of these competitions continue to have an eastern European flavor. The coach of the U.S. Olympiad team in recent years is a case in point. Titu Andreescu, director of the American Mathematics Competitions in Lincoln, Nebraska, is a Romanian émigré. He is a tall, stocky bear of a man with a salt-and-pepper beard and eyes so black they seem bottomless. Titu's mother was born in New York City shortly after her family emigrated from Romania. But she returned to Romania as a child, was raised there, and married a Romanian man. Born in the 1950s, Titu was an avid mathematician as a child and was on the Romanian Olympiad team in 1973. But the government kept close tabs on him. Because his mother had retained her U.S. citizenship, Titu and his mother were viewed as possible defectors. Government officials assumed that if they ever left the country they would not return.

By the late 1980s Titu had become an assistant coach of the Romanian Olympiad team and was editor of a prestigious mathematics journal in Timişoara, Romania's second largest city. But he was restless. Mathematics had given him a way to ignore the numbing repression of life in a Communist country, but outside of his work he felt trapped. The system found a slot for people and kept them in it. In a society grown rigid with dogma, there were few ways to excel.

Suddenly, in late 1989, the Communist bloc began to collapse, and Titu's opportunity arrived. "My mother and I decided to leave," he says. "The government knew that I would eventually defect. I did."

He found a job teaching mathematics at the Illinois Science and Mathematics Academy, a residential high school for academically advanced kids in suburban Chicago. A couple of years later he volunteered to help coach the U.S. Olympiad team in the summers. In 1994, shortly after Titu became an assistant coach, the U.S. team did something that no team from any country had ever done before. At the Olympiad in Hong Kong, all six members of the team received perfect scores on all six problems —

thirty-six perfect solutions. "The journalists wrote about me like I was the Bela Karolyi of high school mathematics," Titu recalls. The next year he became head coach of the team.

Titu is the person who oversees the various tests that culminate in the selection of the U.S. Olympiad team. But he also has a more specific task. Every summer he directs the Mathematical Olympiad Summer Program, a four-week training camp for the six members of the U.S. team and a select group of other high school students who hope to make the team in future years. For kids interested in competitive mathematics, being invited to the summer program — even if you're not on that year's team — is a great honor.

Most summers the program takes place in Lincoln, where the American Mathematics Competitions offices are located and where the students "won't be distracted by a big city," as Titu explains. But the summer before the Forty-second Olympiad, the training camp was instead held at Georgetown University, a few miles up the Potomac River from the Washington Monument. For four weeks the six Olympians and about two dozen other talented high schoolers lived in the Georgetown dorms and worked on mathematics. Between the undergraduates who stay to take summer classes and the high school kids attending camps, a college campus can be a surprisingly busy place in the summer. Amid the aspiring basketball players, ballerinas, and thespians, the math students were fairly inconspicuous — a clump of kids wearing Mathcounts T-shirts making their way to and from the cafeteria.

The group quickly settled into a routine. Each morning and afternoon they congregated in one of the lecture halls on campus. At precisely 9:00 A.M. or 1:00 P.M., Titu or one of the other instructors would stride into the room, usually coming directly from the photocopy machine. Titu sometimes delivered a mini-lecture on a particular subject — combinatorics, for example, or inversions in the plane, or an even more obscure topic. But Titu is not a loquacious person, and his lectures never lasted long. Soon he would pull a sheaf of papers from his valise and pass them through the hall. These were the problems for that session.

The students were not prohibited from working together on the problems. Some preferred to solve as many problems as they could before seeking help on the others, while other students instantly teamed up. Soon the room divided into shifting clusters of bobbing heads as students compared approaches and solutions. Eventually Titu began to make his way up and down the rows, asking students if they were making progress. Sometimes he offered a few words of advice; sometimes he just listened, nodding his head.

After a half hour or so, Titu returned to the front of the room and asked for volunteers to solve the problems. One by one the students went to the board and sketched out their solutions. Though it would be hard to find a more competitive group of high school students, they readily praised one another's work. "That's nice," they'd say. "I like that." Titu was more sparing with his praise. "Good work," he'd say, and the student at the board would blink with surprise.

About halfway through the training camp, a camera crew from the CBS show *Sunday Morning* filmed the students for a couple of days. "Just pretend we're not here," said the producer, even though the cameraman spent the rest of the morning with his lens six inches from the students' faces. But the kids did a pretty good job of carrying on as usual; maybe a generation raised on cable television is not flustered by the idea of their every move being videotaped.

The next day the six team members met with the on-air interviewer, Bob Orr. They endured with good humor the usual goofy questions. "What does a math wizard do to relax?" "What do you think about people who look at guys like you and say, 'Hey, these guys are nerds'?" At one point Orr asked if they knew any math jokes that most people wouldn't understand, and

Gabriel Carroll, a high school senior from Oakland, California, offered the following story.

> Two mathematicians sitting in a restaurant are arguing over whether ordinary people know anything about mathematics. The optimistic one says, "Most people know plenty of math." The pessimistic one says, "No they don't, people are completely ignorant." At that point the pessimist has to go to the bathroom, so the optimist calls their waitress over to the table and says, "When my friend comes back, I'm going to ask you a question, and I want you to reply 'One-third x cubed.' Okay?" The waitress says, "One-third x what?" And he says, "One-third x cubed." So the waitress walks away muttering, "One-third x cubed, one-third x cubed." When the pessimist returns, the optimist says, "I want to prove to you that ordinary people know mathematics." He calls the waitress over and says, "Excuse me, can you tell me the integral of x squared dx?" And she says, "One-third x cubed." Then, as she's walking away, she calls back, "Plus a constant."

Everyone on the team laughed. "I don't get it," said Orr. (For a note on the source of the humor in this joke, see the appendix.)

The *Sunday Morning* segment aired a few days after the conclusion of the Olympiad and was a remarkably good piece of journalism. The students came across as real human beings, not stereotypical nerds. The segment even showed some of the problems on that year's Olympiad. Yet running beneath the narrative like an annoying hum was a constant, unstated question: What kind of person could possibly want to spend four weeks of the summer cooped up in a lecture hall studying math?

∇

Most adults in the United States probably remember very little about the math class they took in the eighth grade. But if you want to relive the experience, here's a sure-fire way to do so. Get the videotape "Eighth-Grade Mathematics Lessons: United States, Japan, Germany" from the U.S. Department of Education and pop it into a videocassette player.

There it is again — an experience almost every adult American has endured. The teacher is asking, "Complementary angles add up to what?" The students are propping up one arm with the other and calling out answers. They stare down at their books or off into space. They appear resigned to serving out their time, as if a punch clock were mounted beside the door.

The tape was made as part of the Third International Mathematics and Science Study, or TIMSS, a massive research project conducted in the mid-1990s. In the fall of 1994 a Los Angeles–based videographer spent seven months driving across the United States, visiting eighty-one junior high and middle schools along the way. At each stop he set up his camera and recorded a single eighth-grade math class. He then dropped each tape into an envelope and mailed it to James Stigler, a professor of psychology at the University of California at Los Angeles.

In a basement conference room at UCLA, Stigler and his colleagues in the psychology department eagerly viewed the tapes as soon as they arrived. They had no idea what to expect. No one had ever tried to videotape a large sample of everyday classes like this. For all they knew, the classes might be so varied that the tapes would yield nothing of value. After all, in the United States local control of education is sacrosanct. Districts and states have different curricula; teachers have different backgrounds; schools serve all kinds of communities, from small towns to inner-city ghettoes. And Americans seem perfectly willing to let teachers maintain a "don't tread on me" philosophy. When teachers go into their classrooms and shut the door, they are free to teach the curriculum more or less as they see fit.

Stigler might have been expecting diversity, but what he found was exactly the opposite. Each taped lesson followed more

or less the same basic script. The teacher usually began by re-
viewing material covered in previous classes, often by checking
homework. He or she then demonstrated the mathematical procedure being studied that day, such as multiplying fractions or
calculating areas. Students were given a worksheet containing
exercises, which they began doing. Near the end of the class,
some of the seatwork was checked and homework was assigned.

What was missing from the classes was almost as obvious as
what was present. Students almost never worked on challenging,
multistep problems. The exercises they did usually required just a
few seconds of routine thought or calculation. In not one of
the eighty-one videotaped eighth-grade math classes did the students work their way through a mathematical proof. Instead,
they memorized formulas and learned how to apply them. Unlike
the Olympians, who use what they know to explore new terri-
tory, the students in the tapes were mostly practicing procedures.

Stigler and his colleagues also arranged for eighth-grade
mathematics classes in Japan to be videotaped, and there they
observed quite a different script. After a short introductory
lecture, the Japanese teacher usually presented a fairly difficult
problem and did not tell the students how to solve it. Students
then worked alone or in small groups to come up with a solution,
while the teacher wandered from group to group to ask questions and offer advice. After ten or fifteen minutes the teacher
called on various students to come to the board and present their
answers to the class. If a student seemed to falter, the teacher offered advice or discussed related mathematical concepts.

This procedure may sound familiar, because it is almost exactly what the Olympians do in their training camp. And for
both the Olympians and the Japanese, the problem-solving approach works. In international comparisons of mathematical
achievement carried out as part of TIMSS, U.S. eighth-graders
were decidedly mediocre, ranking slightly below the average of

the forty-one participating nations. Only seven countries scored significantly lower than the United States: Colombia, Cyprus, Iran, Kuwait, Lithuania, Portugal, and South Africa. Japan and three other Asian countries — Korea, Singapore, and Hong Kong — led the rankings.

For the past several years, Stigler and his colleagues have been analyzing their videotapes at a facility called the Lesson Lab — a converted warehouse on Santa Monica Boulevard just a couple of miles from where old Route 66 runs into the Pacific Ocean. Inside the lab, rows of technicians at bright-colored iMacs translate and transcribe the lessons and store the images and dialogue on the lab's central computer. Then education researchers from around the world pore over the digitized lessons second by second, seeking to find order in the messy, complicated, intensely human act of teaching.

Stigler doesn't blame American math teachers for the lackluster performance of their students. On the contrary, he points out something apparent in all the tapes. The teachers are working extremely hard. They are joking with the students, dealing with interruptions from the public address system, gesticulating at the blackboard, straining to add flavor to material that many of them realize is pretty thin gruel. It seems obvious, after watching the tapes, that teachers' complaints about being exhausted at the end of the day are fully justified.

The problem with U.S. math instruction, says Stigler, lies not in lack of effort by teachers but in the expectations and ingrained behaviors of both teachers and students. The people in a classroom might not realize it, but they are caught up in a cultural activity as rigid and preordained as a religious service. Their behavior is not learned from a book or from teacher education classes. It is learned through experience, starting at a very early age. As Stigler and his colleague James Hiebert point out in their book *The Teaching Gap,* what happens in a classroom is like what happens at a family meal. Everyone has certain expectations about how the food will be served, how conversations will be carried out, and what behaviors are and are not appropriate. People would be startled if Mom handed out a menu at the beginning of the meal or if Dad presented a bill with dessert. Dinner plays out according to a cultural script that everyone knows.

Because teaching is a cultural activity, it is extremely resistant to change, Stigler says. "Cultures represent compromises between the desired and the possible. And the cultural routines that have been created are compromises that have developed over years and years. For that reason, they are overly determined, which means that there's not just one reason why things are the way they are, there are hundreds of reasons why things are the way they are. And even if you can change one of those reasons, the other ninety-nine come back and try to make things the way they were."

Think, for example, about what takes place when an American math teacher asks a simple question, something like, "What is the angle complementary to 84 degrees?" If some student in the class doesn't answer within a few seconds, the teacher starts to get nervous. Teachers in the United States believe that their responsibility is to lead their students in tiny steps, with continual review and reinforcement, from simple counting in grade school all the way to precalculus, calculus, or statistics. If some students can take those steps faster than others, they should be taught separately. But in any math classroom, if students can't answer a simple question involving one of those steps, the chain of understanding is in danger of being broken. The teacher had better go back and cover the material again until everyone has memorized it.

In Japanese classrooms, in contrast, teachers *want* their students to struggle with problems, because they believe that's how students come to really understand mathematical concepts.

Schools do not group students into different ability levels, because the differences among students are seen as a resource that
can broaden the discussion of how to solve a problem. Not all
students will learn the same thing from a lesson; those who are
interested in and talented at math will achieve a different level of
proficiency from their classmates. But each student will learn
more by having to struggle with the problem than by being force-
fed a simple, predigested procedure.

Stigler believes that the only way to change U.S. math education is to reveal the hidden assumptions that constrain students
and teachers. At the Lesson Lab he and his colleagues are compil-
ing an extensive videotape library of actual lessons that have
been analyzed by teachers and education researchers. They want
teachers to use the videos as jumping-off places for what Stigler
calls "lesson study." Groups of teachers would analyze the teaching of a specific task, such as adding fractions. They would do research to understand their students' cognitive abilities and different approaches to teaching that particular task. Then they could
disseminate the results of their study to other teachers.

"The basic idea behind changing a cultural practice is bringing it to awareness," Stigler says. "That's what lesson study does
— it brings to awareness the kinds of practices in which you're
engaged. Once you do that you can start to make some decisions
about whether you want to make a change. But if current practice isn't brought into awareness, when you try to implement a
new practice it gets modified by existing practices in an uncon-
scious way, and you're right back to the way things were."

▽

Titu Andreescu also believes that high-level problem solving is
the key to successful mathematics education — not just for the
best students but for all of them. "In Romania, rather than lec-
turing my students I would choose five or six good problems, and
through those problems students would learn the theory," he

says. "Every teacher in Romania was supposed to be teaching the same lesson at the same time — it was very centralized. But even with those restrictions I succeeded by having a problem-solving approach."

High-level problem solving has many proponents in the United States. Since inaugurating a major reform effort in 1989, the National Council of Teachers of Mathematics (NCTM) has been urging U.S. math teachers to emphasize the kinds of problems that require students to think deeply about what they are doing. But the cultural inertia of math instruction is very powerful. The eighty-one classes Stigler and his colleagues videotaped in 1994 exhibited virtually no signs of the reforms advocated by NCTM.

In some school districts the NCTM's recommendations have sparked intense controversy. Parents and some teachers have assumed that an emphasis on high-level problem solving implies a deemphasis of basic skills like subtracting, multiplying, and dividing. The NCTM insists that this interpretation is mistaken. The "Overview" of its *Principles and Standards for School Mathematics* states, "Students need to learn a new set of mathematics basics that enable them to compute fluently and to solve problems creatively and resourcefully."

The "math wars," as the controversy has come to be called, have helped focus public attention on the deficiencies of U.S. math education, but the deficiencies aren't going away. Given that math teachers already focus largely on "the basics," why do U.S. students rank so low in international comparisons of mathematical performance? Why does math have a reputation as the most dreaded and uninspiring of middle school and high school subjects? How many adults remember their mathematics classes as enjoyable?

"In Romania, everybody involved in math loves math, or at least ninety percent do," says Titu. "In Romania, when they learn that you are a mathematician, they say, 'I was so good at math.' And it is the cab driver who tells you this. 'Math was my favorite subject.' That's one reason why the eastern European teams do so well, because math is part of the culture.

"Here, teachers in elementary school and middle school, many of them hate math. How can you teach math when you hate math? When eight out of ten people in this country learn that I am a mathematician, they say, 'Oh, man, I was terrible at mathematics.'

"In order to change this, you need lots of money. A reporter once asked me, 'If you had a million dollars, how would you change math education?' I said, 'I'll answer your question if you change that *million* to a *billion*.' I think we could do really well at mathematics if the United States invested one billion dollars. But I think it would take that much money to change the attitude of people in this country.

"You need to retrain the teachers, you have to give them an incentive, and if you really want a good program, you have to give them a six-figure salary. For good people in academia to teach in high school, you need to pay them well. If you have a teacher making twenty thousand or thirty thousand dollars a year, you can't expect much from that teacher. To expect quality, you have to pay more.

"Or you have another option. You could buy teachers from different countries. But it's not going to solve the problem. It's going to make it worse, because if we keep importing mathematicians, it will hurt the country eventually. Today a third of the graduate students in mathematics in America are Asians — not Asian Americans, but Asians coming from Asia.

"You also need to get rid of these," Titu said, plucking a thick calculus book from a nearby shelf. "This is a weapon. You can hurt somebody with this. Look." He ruffled through the book. "This has thirteen hundred pages, including appendixes.

One thousand three hundred pages — a calculus book! I learned calculus from a two-hundred-page book, but it was of very high quality. I can't find such a book here. If you open this book at a certain page — look at this, exercise after exercise, look how watered-down this is. Why? Because this book costs eighty-five or ninety dollars. The book I learned from cost a few dollars."

▽

Paul Zeitz, now a professor of mathematics at the University of San Francisco, was a member, along with Eric Lander, of the U.S. Olympiad team that competed in Erfurt, East Germany, in 1974. He's not in touch with his former teammates anymore, though occasionally he hears about them through the grapevine. One, a biostatistician working on environmental issues, lives on an ashram in northern California. Another is a software engineer in Berkeley. A third died in a fall from a tree shortly after returning to the United States from Erfurt.

Zeitz works in a cluttered, windowless office in a building that must have one of the best views in all of America — east to downtown San Francisco, north to the Golden Gate Bridge and Marin County, and west past Golden Gate Park to the all-embracing Pacific. He grew up on the opposite coast, in New York City, where he was captain of the Stuyvesant math team the year after Lander was. He had never been on a plane when he boarded the flight for Germany. He turned sixteen during the trip. "The Vietnam War was going on, and that happened to be the first year that Vietnam came to the IMO," he recalls. "We met their team not long after we got there — some of them could speak French, so we communicated that way. I remember we shook hands. We were all very polite.

"Our coach told us to bring Frisbees, because back then no one outside the United States had ever seen a Frisbee before, and that really worked; we played a lot of Frisbee with the other teams, and you could tell a lot about the other teams by how they played Frisbee. The guys from East Germany just wanted to tackle each other — they thought that was fun. The Russians never did figure out how to throw it — they'd just fling it overhand as far as they could. But the Vietnamese guys, they hung back and watched us for a long time, and when we finally talked them into throwing the Frisbee, they threw it perfectly level. They'd been watching us and learning how to do it. I figure that's why we lost the war."

Zeitz went to Harvard after high school but was turned off by the impersonal teaching of math there. Gradually he got interested in history and decided to major in the subject. But he also did some math tutoring in college and enjoyed it, so after college he applied for jobs as a math teacher around the country. After spending one year at a progressive private school in San Francisco, he took a job at a boarding school in Colorado Springs, where he taught math for five years. "I got a lot of experience there," he recalls. "Private schools are different from public schools. I could invent my own curriculum and do what I wanted."

Zeitz was always interested in applying the problem solving he'd done as an Olympian in his teaching. So he began to develop a series of tough mathematical challenges that his students would work on for extended periods. He found that this approach often engaged the least talented kids just as much as the most talented. "I usually had a class of kids who had been labeled gifted and another class that was essentially considered to be boneheads. And my experience with those classes exactly matched a story I once heard. Two classes like that were asked, 'If you had a bathroom scale and a giraffe, how would you go about weighing him?' The gifted kids got all flustered trying to figure out a way. But one of the kids in the class of boneheads said, 'Well, you'd take a chain saw and cut him up.'

"That's exactly what you need to be good at this kind of

mathematics. The same sorts of things that make some people good problem solvers are the kinds of things that cause kids to have behavioral problems. You have to figure out a way to reward the breakdown of inhibitions. Of course, it has to be done constructively — you don't want anarchy in the classroom. But teachers have to become more comfortable with group activity, with brainstorming, with letting kids make mistakes. You have to go on adventures and not worry about the consequences."

Introducing high-level problem solving into the classroom, Zeitz discovered, helped resolve many of the shortcomings he had encountered in U.S. math education. It kept the talented kids from being bored while offering everyone welcome relief from the endless stream of procedures. It captured the interest of kids who liked to use their brains for more than just memorizing. It justified the need for better teacher training, because teachers need to be able to keep up with their students' mathematical explorations.

Zeitz did a lot of rock climbing and hiking in Colorado, and in the mid-1980s, pursuing a growing interest in geology, he entered a graduate program in geophysics at the University of California at Berkeley. But he found himself ever more attracted to the mathematical aspects of geology, and in 1987 he changed his course of study to mathematics. He got his Ph.D. five years later and immediately landed a job at the University of San Francisco, which was looking for a math professor with teaching experience.

There he began working on *The Art and Craft of Problem Solving*, which was published in 1999. It's a challenging book — many math Olympians use it to prepare for the competition. But Zeitz uses it with all the students in his classes — including those studying to be teachers — often by breaking them up into teams and letting them work on specific problems for days at a time. "The level of the math is not really an issue," he says. "My main goal when I teach this kind of course is to inspire my students to really think hard. It sounds simplistic, but that's all there is to it. I want to get students used to a mindset that a problem that is worth solving is worth pondering for a long time. It's hard to train somebody how to think for a long amount of time. Most people think in maybe five-second bursts. I'm talking about spending days thinking about something. So the level of the math is less an issue than is rewiring their brains so they can concentrate better."

Rewiring the brains of students is an ambitious goal, Zeitz admits. But it can be done by demonstrating that math is not what people think it is. "To be good at math you have to be both rigorous and a good explorer," he says. "And this exploration, this organized play, that's what people don't understand about math — it's a very social activity."

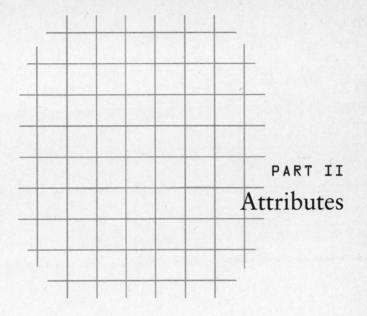

PART II

Attributes

3 . insight

Tiankai Liu was one of the Olympians who was relieved to see the first IMO problem. It was a geometry problem, and Tiankai considered himself a pretty good geometer. Here it is:

> In acute triangle ABC with circumcenter O and altitude AP, angle C is greater than or equal to angle B plus 30 degrees. Prove that angle A plus angle COP is less than 90 degrees.

This problem would not have seemed unfamiliar to the ancient Greeks. The Greeks were the first to show that for every triangle, a circumcircle can be drawn that passes through all three vertices. And the Greeks were the first to think about geometry as an abstract enterprise, separating it from its roots in surveying and measurement. Earlier civilizations had developed considerable geometric knowledge; the Egyptians and Babylonians used the results of the Pythagorean theorem more than a thousand years before Pythagoras lived. But Pythagoras was the first to make that knowledge into a theorem, and in doing so he loosed mathematics from its earthly bounds.

Most people would probably consider this first problem fairly opaque, but as Olympiad problems go, it's straightforward. It starts with a triangle that has vertices labeled A, B, and C. The circumcenter — the center of the circle drawn around the triangle — is labeled O. According to the problem, a single line is drawn from one vertex of the triangle — the one labeled A — to a point called P on the opposite side of the triangle. The rest of the problem simply involves the sizes of the angles defined by the five points A, B, C, O, and P.

The first thing Tiankai needed to figure out was how the five points fit together. So he took the compass and straightedge that the Olympians are allowed to bring to the test and drew the following diagram, which includes all the pieces of the problem.

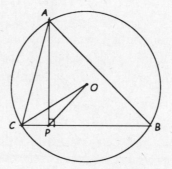

The "altitude AP" is the line starting at A that extends to point P on the line CB. It's called an altitude because it meets line CB at a right angle (the little square at the base of line AP indicates that angle APB is 90 degrees). Angles A, B, and C are simply the three angles that make up the triangle. (In other words, the angle ABC can simply be called angle B, a shorthand that Tiankai used throughout his solution.) The problem says that angle C (or BCA) is at least 30 degrees larger than angle B. Tiankai needed to prove that angle A plus angle COP is less than 90 degrees.

One of the tricks in solving Olympiad problems is not to let your mind get locked into a potential solution too quickly, because you can waste a lot of time going in the wrong direction. So Tiankai let his thoughts wander. He continued line CO to the opposite side of the circle and looked at the angle BCO, but that didn't seem to take him anywhere. Then he drew a line starting at B and passing through O to the opposite side of the circle. It made for a pretty diagram, but that approach also didn't seem to hold much promise. He needed some sort of insight, a way of viewing the problem that would reveal how to solve it.

"The thing I was motivated by was that thirty-degree number," says Tiankai. "I knew I had to do something with that, but I didn't know what. So I just sat there for a while. I wasn't nervous. I knew I'd come up with something."

▽

Tiankai's calmness was one of the first things an observer would note about him — not because anyone would have expected him to act otherwise, but because he was by far the youngest member of the team. The other five were all high school graduates who would be going to college in the fall. Tiankai had just finished his freshman year at Phillips Exeter Academy in New Hampshire. A foot shorter than most of his teammates, he was so skinny and small that he looked like someone's kid brother. When the Olympians from other countries saw the U.S. team walking together, they often asked, "Is that kid a member of your team?" When told that he was, they shook their heads in disbelief.

Tiankai tried to ignore comments about his age. Though not one to back down from any challenge, he tends to deflect attention from himself. He often holds his body a bit sideways, as if the world strikes him as slightly absurd and he's watching it with amusement. Though proud of his mathematical abilities, he doesn't want to give the impression that he's working too hard. Sometimes he erects a psychological shield around himself by harshly belittling his achievements. "I used to like chess a lot, but I wasn't particularly good at it — I was like fifteenth in some state competition in elementary school," he says. "I also did some computer programming, but maybe that's because both my parents are computer software engineers. And I play the piano. I wish I wasn't so lazy and practiced more, but still I like it."

Tiankai was born in Shijiazhuang, a city about two hundred miles southwest of Beijing. Like Melanie, he displayed a remarkable interest in mathematical concepts from a very early age. "We used to take a long walk after dinner when Tiankai was about one and a half years old," recalls his mother, Kaining Gu, "and Tiankai was extremely interested in the sizes and shapes of the manhole covers on the ground. When he discovered a new cover, he would become very excited. His eyes would open wide, his breath would get short, and he'd call out in Chinese: 'Circular cover, square cover, big cover, small cover, big circular cover.'"

In 1989 Tiankai and his father left their homeland to join Tiankai's mother in Darmstadt, West Germany, where she was taking graduate courses. A year later the family immigrated to the United States and settled in San Jose, California, where Tiankai's parents both found work as software engineers. Throughout this period Tiankai continued to be fascinated by mathematics and by games with a mathematical edge. "His random drawings were usually full of a matrix of numbers," his mother says. "At his fourth birthday, a friend of ours sent him a chess set, and he soon learned how to play. That time we were in Darmstadt, and we used to take a walk to a nearby park to watch people playing chess. A chess coach in the town once wrote an article about Tiankai. He was impressed that at such a young age Tiankai had learned how to plan attacks."

In San Jose, Tiankai attended the public schools and took more or less the same math classes as everyone else his age. But his abilities soon attracted attention. His mother remembers that once when Tiankai was in kindergarten, as they were walking to the park, some older students in his school stopped and asked him some math questions, which he answered. She also heard that he was pointing out mistakes made by his teachers in class or in his textbooks. "Often during the parent-teacher conferences his teachers would ask us if his math capability was by nature or by learning. They would ask me if he had any tutoring at home. The fact is that we have never paid anyone to teach him math. He loves it and he learned a lot of it by himself. I'd say it's a mixture or interaction of the two ingredients — one from nature, the other from learning."

Tiankai attributes at least part of his interest in math to another factor. Like every other member of the U.S. Olympiad team, Tiankai speaks perfectly inflected, slang-filled English. But in elementary school he felt less confident about his English. At home he speaks Chinese with his parents, and he still reads many books in Chinese. "I wasn't superduper in English, partly because my parents didn't know English very well," he says. "So I decided that maybe I could do math."

In 1998, when Tiankai was in seventh grade, his parents read an article in the *San Jose Mercury News* about a group of mathematicians who were organizing a series of math circles in the San Francisco Bay area. Math circles are another import from eastern Europe and Russia, where for more than a century mathematics professors have organized afterschool math groups for nearby secondary school students. This tradition is one of the reasons for the rich history of problem solving in those countries. Math circles also contributed to the founding of some of the high schools specializing in math and science that still train many eastern European and Russian Olympians.

The Bay Area math circles had several organizers. One was Paul Zeitz, who was then just finishing up *The Art and Craft of Problem Solving*. Two others were Tom Davis, who cofounded Silicon Graphics, and Brian Conrey, executive director of the American Institute of Mathematics in Palo Alto, a privately funded organization that supports cutting-edge mathematical research. And the most active organizers were two female mathematicians: Zvezdelina Stankova, of Mills College and the University of California at Berkeley, who was on the Bulgarian Olympiad team in 1987 and 1988, and Tatiana Shubin of San Jose State University.

"I was brought up and educated in the old Soviet Union," says Shubin, "so I had a different perspective on mathematics. When I was growing up, everybody knew that the smartest kids on the block were doing mathematics, and we were very well respected. We weren't math freaks, we were smart kids, and even people who weren't interested in mathematics would respect us. It was very different when I came to this country. Whenever someone asked what I was doing and I said mathematics, people would immediately shrink from me, as if I had said something unpleasant. I couldn't understand that.

"And then when my daughter started school I had another shock. I remember when she was in maybe the third grade I was looking at her homework problems, and they were just painting bubbles rather than doing mathematics. I went to the principal to talk about this, and the principal was absolutely appalled when I told him that the kids were not using parentheses properly in mathematical expressions. It turns out there was no such thing as mathematical expressions as far as this principal was concerned.

"From that point on I decided I wanted to do something for the kids in this country, and especially for the bright kids, because I noticed that the system was going out of its way to help underachievers and was doing nothing to help overachievers."

The first San Jose math circle was held in November 1998. "We had about thirty kids," says Shubin. "We had no idea where they came from, because we didn't screen them. Zvezda gave that first lecture, and she explained inversion in the plane, which isn't a trivial thing for kids of that age, and then she gave them some problems and maybe five minutes at the most to work on them. Then she asked whether anyone would be willing to show a solution. And this little guy — he looked much younger than he was at the time, he was very small for his age — went up to the board and showed his solution, and it was a real solution. There was nothing wrong with it, and he presented it in a very nice and concise form. It was Tiankai, of course. That was when we knew we had some very good kids there.

"The next lecture was on elementary number theory, a completely different subject, and it was given by Brian Conrey. He started by introducing lots of notation. I've taught number theory here at San Jose State several times, and I know that to swallow all that notation and get used to it takes time, even for students at the college level. Brian asked if any of the kids had seen that notation before, and none of them said they had, so I'm sure they were seeing it for the first time in their lives. Then Brian introduced Fibonacci numbers and explained what they were, and he asked the students to prove that the fifth Fibonacci number divides every fifth Fibonacci number after that.* He gave them two or three minutes and then asked whether anyone had solved the problem. Someone from the very back of the room, very quietly, but loud enough for people to hear, said, 'I think it can be generalized.' Brian asked, 'In what form?' And this boy said, 'I think the mth Fibonacci number, for any m, divides the (m times n)th Fibonacci number, for any n.' Brian said, 'Well, how would we go about proving something like that? Let's try some examples.' But he was really just dragging his feet because *he* was trying to solve this problem. And then Tiankai — it was Tiankai, of course — said, 'Yes, but I can prove it.' And he marched to the board and proved it beautifully, with no errors. From that day on we knew we had a real star in that class. There wasn't a single math circle where he didn't do something unusual or hard.

"We also had, together with these circles, a series of lectures given by first-class mathematicians for bright high school

---

*Fibonacci numbers are generated by adding the two previous numbers in a particular sequence to get the next number. By convention the first two Fibonacci numbers are 1, so the third is 2; F_4 is 3, F_5 is 5, F_6 is 8, and so on. Conrey was asking the students to prove that F_5 evenly divides F_{10}, F_{15}, F_{20}, and so on.

students. We had three in the fall and three in the spring, and Tiankai was at each of those lectures. At the very end of the 1998–99 school year an additional lecture was given by Andrew Wiles, who solved Fermat's last theorem. It was hosted by the American Institute of Mathematics and the mathematics department of Stanford University and was held at Stanford, in a big auditorium that seats more than seven hundred people. All of the students involved with the math circles got free tickets, but there were also lots of adults from all over the Bay Area — I recognized people from Berkeley and even farther away. The auditorium was absolutely packed. At the end of the lecture two microphones were installed in the aisles, and people were told that if they had a question they should stand in line at one of these microphones and they would be given a chance to ask their question. I was more than a little surprised to see Tiankai go to one of those microphones. He was the last person to ask a question, because I think he was shy and didn't want to get in front of any adult. But he patiently stood at the end of this huge line, and his was the most interesting question asked that entire evening."

Tiankai participated in Mathcounts in the seventh grade and finished fifth in the nation. The next year, as an eighth-grader, he finished first in the written portion of the test but got knocked back to second during the head-to-head competition. Eighth grade also was important for Tiankai because he left the public school system that year and entered a private school in the San Jose area. His experience there was "extremely positive," in his mother's words. After two weeks Tiankai's math teacher decided that he was wasting time in geometry and moved him into a precalculus class for high school students. The school also encouraged Tiankai to speak up more, both in class and outside. "He's usually a quiet boy," his mother says.

Tiankai and his parents began thinking about private high schools. "Tiankai applied to Andover and Exeter," his mother says. "We liked the rich history and tradition of the schools and their strong English and humanities courses — those are what Tiankai needed most growing up in an immigrant family." He was accepted by both. Then the family learned that one of the math teachers at Exeter, Zuming Feng, who also had immigrated from China, was a coach for the U.S. Olympiad team. It seemed the perfect match, and late in the summer after eighth grade, Tiankai boarded a jet bound for Boston.

▽

Mathematicians have always been fascinated by accounts of precocious mathematical achievements. They all know the story of Carl Friedrich Gauss, who was born in Brunswick, Germany, in 1777. When Gauss was three, his father was making out a weekly payroll when the little boy, peering over his shoulder, corrected his addition. When Gauss was ten, the teacher at his school decided to keep the students busy by having them add the numbers from 1 to 100. Gauss had never seen the problem before, but he immediately figured out a clever way to calculate the sum quickly. He wrote the answer on his slate, marched to the front of the room, and deposited the slate on his teacher's desk. Later in life Gauss liked to recount how his was the only correct answer, even though his classmates worked for hours laboriously adding number after number.

This fascination with precocity has subtle but pervasive effects throughout mathematics. The equivalent of the Nobel Prize in mathematics is called the Fields Medal, but unlike the Nobel it is given only to mathematicians aged forty or younger. One of the things that may have sparked John Nash's schizophrenia — so ably documented in Sylvia Nassar's book *A Beautiful Mind* — was his fear as he approached his thirtieth birthday that his best work might be behind him. An often-quoted passage from *A Mathematician's Apology*, by the English number theorist G. H. Hardy, involves age: "No mathematician should ever allow himself to forget that mathematics, more than any other art or science, is a young man's game." As the French mathematician André Weil once wrote, "There are examples to show that in mathematics an old person can do useful work, even inspired work; but they are rare and each case fills us with wonder and admiration."

Actually, a careful study of the mathematics literature disproves this stereotype. In her article "Age and Achievement in Mathematics: A Case-Study in the Sociology of Science," Hofstra University historian Nancy Stern found that mathematicians over thirty-five publish just as many papers as do mathematicians under that age. And the papers of older mathematicians are cited more often than those of younger mathematicians, suggesting no decline in quality.

Women in the profession point out that the myth of the young hotshot mathematician is especially damaging to females. During their twenties, when most universities and employers would just as soon have their mathematicians doing nothing but math twenty-four hours a day, women with families have other responsibilities. Many women say that they've done their best work later in life, after their children were grown. And they point to many famous mathematicians who did superlative work their whole lives.

Nevertheless, mathematicians continue to marvel at young people who exhibit profound mathematical ability, and this tendency can be especially strong at an Olympiad. That someone as young as Tiankai can solve such difficult problems doesn't seem possible — or entirely fair. No one gets to the Olympiad without working extremely hard. Very young people seem not to have paid their dues.

Then again, another theory holds that youth can sometimes be an advantage at the Olympiad. According to this line of thought, young minds aren't yet cluttered with all kinds of mathematical debris and can cut right to the heart of a problem. "Younger students rely a lot on intuition because they don't have as much knowledge, and they don't necessarily do as much analysis of what they do," says Alex Saltman, who was on the U.S. Olympiad team that went to India in 1996 and who has taught at the summer program for the past several years. Young mathematicians may not have the sophistication to figure out whether their intuitions are accurate, but they can spin out ideas and conjectures at a ferocious pace.

This theory has a corollary, which is that younger mathematicians often have a particular skill that makes up for their lack of experience. Many seem to have an uncanny ability to picture mathematical problems in their mind. Their inner vision — literally, in-sight — enables them to manipulate a problem as if it were an object lying before them on a table. They can see things in their heads that other people cannot, and these images suggest ideas that do not occur to the rest of us.

Some of the world's great creative geniuses had this skill. Albert Einstein once said he did not think using words at all. Rather, he thought in terms of signs and images, which he could move around and recombine at will. Einstein conceived of his special theory of relativity by imagining himself traveling at the speed of light next to a light wave. He developed his general theory of relativity by picturing a man in a box falling down an endless shaft.

Other important scientific and mathematical advances originated as vivid mental images. The English physicist Michael Faraday developed his theories of electric and magnetic fields after envisioning invisible tubes arcing through space. James Watt conceived of the steam engine while walking in the country one day and picturing in his mind how a cylinder could be connected to a condenser. And many great creators outside of science have had intense powers of visualization. Michaelangelo reportedly could remember every detail of every piece of art he had ever seen.

Roger Shepard, who is now a Stanford University professor emeritus, has been interested in mental images for as long as he can remember. In his book *Mind Sights: Original Visual Illusions, Ambiguities, and Other Anomalies, with a Commentary on the Play of Mind in Perception and Art,* he writes:

> Beginning in childhood, I vented through endless drawings what one of my despairing elementary school teachers termed my "feverish imagination." At one stage, my insatiable requirements for paper drove me to obtain the unused ends of rolls of newsprint from the publisher of the local daily paper. These early drawings typically portrayed vast, other-worldly vistas whose desolation was here and there broken by strange, solitary towers, futuristic vehicles, machines, robots, and much later (and to the extent that my skills grew adequate to the task) beautiful women.

During his career, at Harvard, Bell Labs, and Stanford, Shepard did pioneering research on how people perceive objects, sounds, and music — work for which he received the National Medal of Science in 1995. Yet he kept drawing, fascinated by images that exist in the mind but not in the real world. He drew the shifting patterns that appear when one applies pressure to the eyeballs. He was especially interested in the images that sometimes flash vividly in the mind when one is falling asleep (called hypnagogic images) or when waking up (hypnopompic images).

One of Shepard's hypnopompic images is now familiar to many millions of people. As he was waking up on the morning of November 16, 1968, he saw in his mind "a spontaneous kinetic image of three-dimensional structures majestically turning in space." He immediately grabbed a bedside pad and jotted down an idea for an experiment that would clarify an especially thorny problem in cognitive psychology.

The image Shepard saw that morning has since been converted into a widely used test of mental agility. Here's a sample item from such a test; your task is to decide which of the objects shown below are identical to the circled object.

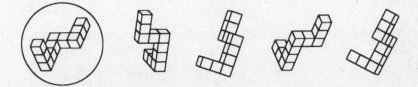

Shepard's original experiment, the results of which were published in *Science* magazine in 1971, explored how people make these comparisons. Shepard and his coauthor, Jacqueline Metzler, showed sets of three-dimensional block structures to people and asked them to determine whether the objects were the same. They found that the amount of time a person needed to answer the question depended on the extent to which an object had been rotated. Apparently, people were rotating the objects in their minds to see if they corresponded with the other objects. Shepard and Metzler were able to show that the average person could mentally rotate an object at about 60 degrees per second.

These results were a shock to many cognitive psychologists. At the time, most of Shepard's colleagues thought that people would make such comparisons in a completely different way. The reigning view was that mental images were structured the same way as language. We might believe we "see" something in our minds, but the mental representation of that object was actually just a list of verbally encoded attributes. The image was an illusion, a trick. Some psychologists went so far as to claim that mental images do not exist. After all, where in our minds does the image appear, and who is there to see it?

Shepard's experiment demolished this argument. If the objects in Shepard and Metzler's experiment were really just lists of verbal attributes, then the ability to compare two objects would not depend on how much they had been rotated. People would simply compare the attributes of the objects and decide if they were the same. But people weren't thinking in words while doing a mental rotation test. They were using a different kind of language, a language of shapes and movements.

▽

Mathematicians and artists are not the only ones who use mental images to solve problems. People use visualization all the time — to rearrange the furniture in a room, to decide whether a car will fit into a parking space, to estimate the amount of bunting to buy for a party, to figure out whether one-fourth or two-fifths is the larger number. But not everyone uses mental imagery to the same extent or in the same way.

At Boston College, psychologist Beth Casey and her colleagues have developed a simple test to determine a person's preferences in processing information. Look at the following phrase:

Triangle above circle

Now turn to page 74 and — without looking back at this page — decide whether the image drawn there matches the above phrase. Now think about how you decided that question.

1. Did you memorize the words in the phrase, turn the page, and then compare the memorized words to the picture?
2. Or did you convert the phrase into a picture in your mind, turn the page, and compare the image in your mind to the picture on the page?

Casey and her colleagues have found that about half the people they test use the first strategy — they tend to rely relatively little on mental images. They remember the words, look at the image, convert the image into words, and see if the sets of words are the same. This first group Casey calls verbalizers. The second half uses the second strategy or some combination of the two. They convert the words into an imagined image and then compare their mental picture with the real image. These people are, in Casey's terminology, visualizers.

Verbalizers and visualizers solve problems differently, Casey says. For example, Casey considers herself a verbalizer. With an image from a mental rotation test, she would not imagine the entire object and rotate it through space. Instead, she would rely on her logical and analytic skills to solve the problem. "I'd imagine myself standing right here on this block," she says, pointing to a block in the middle of the object, "and then I'd look around me and see what I'd see. So I'm not really rotating the image. That's the problem with most tests. People are always able to figure out a way to bypass what the test is trying to measure."

Casey's husband, on the other hand, is an intense visualizer. In fact, when his brain is working on an image, she says, he can't be bothered to talk. "If you ask him a question when he's doing one of these tests, he won't answer, because he says that it breaks up the image in his head. In his mind, these two processes are competitive. He can't visualize and verbalize at the same time."

Girls often test lower than boys on measures of spatial ability, but Casey believes that anyone — male, female, young, old — can learn to be a better visualizer. (The role of genetics versus experience in this gender difference is another focus of Casey's research.) In turn, better visualization abilities can give anyone a step up not only in math classes but in a wide range of daily tasks. "For example, the idea that one-fourth is larger than one-eighth is a very difficult idea for young students, because the number eight is larger than the number four. But if a student can visualize what a pizza slice would look like if you had one-fourth or one-eighth of that pizza, that would help. People who don't have that ability can have problems with fractions their whole lives."

For the past several years, Casey and her colleagues Ronald

Nuttall and Elizabeth Pezaris have been developing a series of story-based lessons for young children that promote the development of spatial skills. In the story *Tan and the Shape Changer*, children combine triangles to make more complex shapes. In *Sneeze the Dragon Builds a Castle*, they build model castles using blocks, culminating in the construction of arches, bridges, and towers. "The idea is to use the story to make spatial thinking a meaningful activity and therefore a way of approaching math problems from the very beginning," she says.

Better curricula and teaching won't turn everyone into a visualizer, Casey acknowledges. But careful instructions can help people develop the same skills that many mathematicians use to solve problems. Of course, most people won't apply their visualization skills to the same ends as the math Olympians do. And the Olympians have an additional strategy, Casey believes, that contributes to their dexterity as problem solvers. They are able to look at problems from different and unconventional perspectives. "I've known a lot of mathematicians, and I think it's their atypical way of thinking that makes them different," she says. "They can take information and come at it in a novel way, either spatially or analytically. And the truly gifted mathematicians are the ones who are good at both spatial and verbal representations."

▼

As a mathematician, Tiankai considers himself both a verbalizer and a visualizer. But the skill that enabled him to solve the first Olympiad problem was his ability to see the problem as an image, an integrated whole, rather than as disconnected lines and a circle drawn on a page.

After looking at his initial sketch for a few minutes, Tiankai realized that he needed a few more points and lines to anchor his work. So he drew lines from the circumcenter, O, perpendicular to lines AP and CB. He labeled the ends of these perpendicular lines X and M, respectively. He then drew in a few other useful lines, so that his diagram now looked like this:

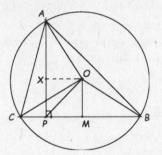

Tiankai's task was to prove that angle A plus angle COP is less than 90 degrees, given that angle C is at least 30 degrees more than angle B. Point P is the critical factor, he quickly realized. As P moves closer to point M, angle COP gets bigger. But if angle COP is too big, it cannot be added to angle A and still be less than 90 degrees. He had to prove that P was sufficiently far from M for the condition to hold.

So what controls the location of point P? It's angle PAO, Tiankai soon saw. If he could figure out the size of that angle, he would have a good chance of finding out where point P is situated. But how to determine angle PAO?

He gazed at the problem for a few minutes and then thought of something. He could calculate the size of angle CAO and then subtract angle CAP. What was left would be angle PAO.

Figuring out angle CAO was not terribly difficult. Tiankai knew that the angles AOC and ABC are related in a particular way. The two triangles have two points in common, C and A. Also, point O is the center of the circumcircle and B is on the circumference of the circumcircle. According to a classic theorem in geometry, angle AOC is twice the measure of angle B.

But now look at triangle CAO. Two of its sides are the same length, because they are both radii of the circumcircle. Thus CAO is an isosceles triangle, which means that two of its angles are the same. The angles of any triangle add up to 180 degrees, so a little algebra is sufficient to show that angle CAO is equal to 90 degrees minus angle B. Tiankai was halfway to his goal.

Now he had to figure out the size of angle CAP. That's even easier. One angle of the triangle formed by points C, A, and P is 90 degrees, and the other angle is ACP, which is the same as angle C. So angle CAP is equal to 90 degrees minus angle C.

At this point Tiankai could calculate angle PAO by subtracting $90 - C$ (the measure of angle CAP) from $90 - B$ (the measure of CAO). The two 90s cancel out, and what's left is the following: angle PAO $= C - B$.

Now that's a remarkable result. How can one tell, by looking at Tiankai's diagram, that angle PAO is so strictly determined by the triangle within which it's embedded? This is the crux of the solution, because the problem says that "angle C is greater than or equal to angle B plus 30 degrees." Therefore, angle PAO has to be greater than or equal to 30 degrees. "That was very helpful," says Tiankai. "You want that to be more than 30 degrees."

But Tiankai wasn't done yet, because he still had to figure out the size of angle COP. To do that, he began to focus on the distance from C to P and from P to M. The closer point P is to C, the smaller angle COP becomes. The closer P is to M, the larger the angle. How could Tiankai peg the size of these angles?

At this point Tiankai used one of the most elegant tricks in all of mathematics. First he assumed the opposite of what he wanted to show. Then he proved that this assumption led to an impossibility, so that what he wanted to show in the first place had to be correct. It's called arguing by contradiction. Say, for example, that you want to demonstrate to a friend the impossibility of walking through a brick wall. You could assume the opposite, stride forcefully toward the wall, and bloody your nose against it. The proof is complete: you can't walk through a solid brick wall.

The details of Tiankai's proof by contradiction are described in the appendix (though ambitious readers might want to try working out this part of the proof on their own). Essentially, he assumed that P is closer to M than to C — in other words, that the distance CP is greater than PM. But given that angle PAO is more than 30 degrees, this leads to an impossibility. So Tiankai's assumption had to be wrong; P has to be closer to C than to M.

Now he was almost, but not quite, done. If P is closer to C than to M, angle COP has to be smaller than angle OCP. (Looking at the triangle for a few moments will show you why this is so.) But angle OCP is part of the isosceles triangle OCB. As was the case with angle B, angle COB is equal to two times angle A. The same algebraic manipulation used previously demonstrates that angle OCP is therefore equal to 90 degrees minus angle A. But Tiankai had shown that angle COP is smaller than angle OCP. Therefore angle COP has to be less than 90 degrees minus angle A, because that's the measure of angle OCP. Tiankai was done — the problem had asked the Olympians to "prove that angle A plus angle COP is less than 90 degrees."

Tiankai solved the first Olympiad problem in about forty-five minutes. He had more than three hours to work on the next two problems. But the first problem was traditionally the easiest, he knew. The hard parts were still to come.

▽

To those who have forgotten most of the mathematics they learned in school, the above proof might seem fairly involved. But to math students in high school or college or to scientists or engineers, the proof may not seem that difficult. As mathematicians often observe, once a proof has been explained it can seem trivial.

But solving an Olympiad problem is never trivial. Zuming Feng, Tiankai's math teacher at Exeter, puts it this way: "It's like you watch television, you see those football kickers, they make a thirty-yard field goal, and you say, 'Yeah, that's easy.' But if we go down there and try that, ninety-nine percent of us will not make it."

The constant ticking of the clock is part of the challenge. Even if the Olympians solve one problem, they know they have two more to go that day. For that reason they often don't solve the problems in the order they're given. Some may work on all three problems for four hours and only begin to stumble across solutions in the final half hour.

An even greater challenge is deciding how to approach a problem. "For most problems there are lots of approaches, but you don't know which one is correct," says Feng. "So if your approach is not working, you face a huge dilemma. Do you want to try to fix your approach, or do you want to start again? This is a very hard choice to make, because you never know how close you are to a solution. When do you want to make that change? It's very hard to decide."

This first Olympiad problem can be solved in several ways. But none of the proofs turned in at the Olympiad were simpler or more elegant than the one Tiankai produced. The wildly prolific and eccentric twentieth-century mathematician Paul Erdős often talked about an object he called the Book, which, he said, was God's collection of the most beautiful proofs of all possible theorems. No mathematician can be absolutely sure that a given proof is in the Book, since there may always be an easier or clearer way to solve a problem. But of all the solutions turned in for this first problem at the Olympiad, Tiankai's is closest to the Book proof.

And it is the very elegance of Tiankai's proof that raises the final, and perhaps unanswerable, question. How did he manage to develop what might be the Book proof in just forty-five minutes? How did he know so quickly and so surely which angles to draw, which points to add, which lines to compare? He can't explain. When you ask him, he shrugs his shoulders and turns away.

4. competitiveness

Under the lights on the Georgetown University football field, the kids at the Olympiad training camp were playing Ultimate Frisbee. The two teams lined up on opposite goal lines, and one of the kids sent a seventy-five-yard pull spinning down the pitch. Oaz Nir fielded the Frisbee at the twenty-yard line and looked toward his receivers, who were feinting and cutting to get free. It was a hot, humid evening in Washington, and the air was completely still, even though the football field was perched on a hill overlooking the Potomac. The players began to sweat, and the lights high above the field reflected off their glistening skin. Most of the thirty or so kids at the training camp were in the game, and while some were obviously not very athletic they played enthusiastically. Others — including several members of the Olympiad team — were in excellent shape, with lean, muscular physiques; they pivoted and leapt with the grace that comes from spending many hours at sports.

Ultimate Frisbee was invented in 1968 by a group of students attending Columbia High School in Maplewood, New Jersey. The school newspaper staff and student council members decided to form a school Frisbee team, partly to capitalize on the sudden popularity of the toy, partly to poke fun at the school's more conventional sports teams. At first the team attracted mainly the more academic kids and the druggies, and the games they played were fairly free-form. But over time other kids began to play, and the rules slowly took shape. Whoever catches the disk has to keep one foot stationary, just as a player who stops dribbling in a basketball game does. That player then has ten seconds to throw to another player on the team. The opposite team can intercept the Frisbee or, if it hits the ground, the other team takes possession. Each team scores by advancing the disk past its goal line.

The first interscholastic Ultimate Frisbee game was played on November 7, 1970, when nearby Millburn High School sent a team to play in the Columbia High School parking lot. Other schools in the area began asking for copies of the rules, and five high school teams formed the New Jersey Frisbee Conference in the spring of 1971. Meanwhile, New Jersey high school graduates were taking the game with them to college, and the first intercollegiate tournament was held at Yale in 1975.

In 1980 forty Ultimate organizers met in Atlanta to discuss setting up an international organization for their sport. A world association was created in 1984, and the first World Flying Disc Federation Congress took place in Sweden in 1985. Today more than one hundred thousand people play the game in more than forty-two countries, and it was a medal sport in the 2001 World Games in Japan.

The rules of Ultimate Frisbee feature a careful mix of competition and cooperation. Players on the field decide whether a score has occurred and call their own fouls (an early definition of a foul was "any action sufficient to arouse the ire of your opponent"). If a call is disputed, the play is rerun. Over the years the intent of the rules has coalesced into a body of knowledge known as the "spirit of the game." According to the official rules, "Such actions as taunting of opposing players, dangerous aggression, intentional fouling, or other 'win-at-all-costs' behavior are contrary to the spirit of the game and must be avoided by all players." Even in the world championships there are no formal officials, and the Spirit of the Game award given to a player or a team at each tournament is highly coveted.

Competitiveness, cooperation, aggressiveness, perseverance — motivational factors are as important in an International Mathematical Olympiad as in any athletic endeavor. And this motivation arises from a similarly complex mixture of psychological forces: the desire to do well, not to be embarrassed, to beat the other guy, to enjoy the game, to sweat, to garner praise, to win. The Olympians do math problems with the fervor and single-mindedness of world-class athletes. Why? What combination of aptitudes and experiences accounts for their fierce competitiveness?

▽

Ian Le had been the final person selected to be a member of the U.S. Olympiad team. Membership on the team is determined by two tests. The first, the USAMO, held in May, selects the twelve high schoolers who can compete for the six slots on the team. The second test, the qualifying exam, is taken by the students at the summer training camp during the first few days of the program.

The morning after the qualifying exam for the Forty-second Olympiad, everyone gathered in the main lecture hall to hear who would be on the team. A few of the two dozen or so students in the hall were shoo-ins — they had been on the team in previous years and had done extremely well on the USAMO. Many of the younger kids, on the other hand, knew they had no chance; they were at the training camp mainly to improve their odds for future years. Still, that left a sizable group in the middle who could hope to make the team — if their scores on the qualifying exam were high enough.

At 9:00 A.M. Titu strode into the lecture hall. He pulled a list from his valise and began to read. Reid Barton, Oaz Nir, and Gabriel Carroll would be on the team — no surprises there, since all three had gone to the Olympiad in Korea the previous year. The next two names Titu called out were Tiankai Liu and David Shin; the only real surprise was that Tiankai would be on the team as a ninth-grader. For the sixth spot, there was a three-way tie, Titu announced. A special run-off exam would be held that afternoon to pick the final team member.

At that point it should have been obvious to everyone that Ian Le was going to be very hard to beat. The qualifying exams had become a test of endurance as well as mathematical prowess, and no one at the training camp was tougher or more determined than Ian. Since elementary school he had been a swimmer, adhering to the rigorous, self-imposed regime of those who steep themselves in chlorine — two hours a day in the pool before school, sometimes another practice in the afternoon. He had the thick, triangular torso of a backstroker rather than the lean taper of a freestyler or breaststroker, and he carried himself with a languid grace. He was also an accomplished pianist who had given several solo recitals. "I never practiced the piano as much as I should," he said. "My teacher always wanted me to practice an hour and a half each day. If I was lucky, during the week, I'd practice maybe forty-five minutes a day, and then maybe an hour and a half on Saturday and on Sunday."

Ian was born in Australia in 1983 to parents who had emigrated from Vietnam during the war. His father, Tri Le, was born in what is now North Vietnam, but the family moved to Saigon in 1954 after the Communist takeover of the north. There Ian's father grew up desperately poor but an excellent student.

"I came to Australia in 1971 under a program called the Colombo Plan," Tri Le says. "It offered scholarships to very good high school students, mainly from Southeast Asia, but some from as far away as Africa. At that time, if you were doing well in school, you could postpone your military service. But as soon as you finished college, you would be drafted into the army. I was just about to finish college when the war ended. That was my senior year when Vietnam fell."

In Australia Le met a younger Vietnamese woman in the same program, who would later become his wife. When the war ended and the Communists took over, Tri Le and his future wife were still in Australia, but their parents and siblings were in Vietnam. Le's brother-in-law was a helicopter pilot and a captain in the South Vietnamese Air Force. "He was based in Bien Hoa," says Le, "which is about twenty miles from Saigon. The day Saigon fell he began flying family members out to the Seventh Fleet. He also picked up other people, not just our families, but other people who wanted to leave. What happened is that normally you flew out to the fleet and they kept the people there and abandoned the helicopter. But we have a big family on my side and on his side, and he made quite a few trips back and forth. He got some of his family out and most of my family out.

"Then, on his last trip to pick up his parents and my parents and my two younger brothers, a lot of people knew about the situation and tried to get on the helicopter. You've probably seen the pictures of the helicopter that was taking off from the American embassy with people hanging on — it was like that. He took off, and a lot of people were hanging onto the helicopter, and it was just too heavy. Somehow the helicopter lost its balance, so the propellers hit an electrical pole, and it went down. He was stuck behind."

In 1981 the Les visited Vietnam from Australia to see their families. "Things were very bad, everyone was struggling," Le says. "Six months before we arrived, my brother-in-law had been released from the reeducation camp, or I guess you would say the concentration camp. He tried to escape from the country again and was caught and put in jail. So when we went to Vietnam he was in jail and we never did see him.

"Eventually they allowed him to go to the United States to be united with his family, in 1990. So he was stuck behind for fifteen years."

Tri Le, his wife, and their two young children immigrated to the United States in 1991. Tri Le found a job as an engineer in Philadelphia. His wife went to work as an investment analyst in New York City. To split the difference between their commutes and to live in a good public school district, the Les moved to Princeton Junction, New Jersey, where Ian, their oldest son, entered second grade. In kindergarten back in Australia, Ian had been good enough in mathematics that his teacher had given him different problems from those his classmates received. But in elementary school in Princeton he did pretty much the same math as everyone else in his class. In fourth grade he entered a gifted and talented program with a few other students, but he wasn't doing anything that thousands of other bright children throughout the United States don't do.

The turning point came in eighth grade, when a teacher in his middle school introduced him to Mathcounts. He did well at the chapter level and then was one of the top four finishers in New Jersey, so he made the state team. At the nationals he finished twenty-sixth out of the 228 competitors. "I had really wanted to place in the top ten," Ian says. "I was kind of upset."

Ian was hooked. He bought several books of Olympiad-level problems and began working his way through them. Sometimes he would struggle over a single problem for hours, but he told himself that it was good training. He took a math class at Princeton, and he read his father's old college textbooks. His problem solving was improving dramatically, yet still he wondered if he could aspire to the top levels of high school mathematics. "I didn't think I could make it to the summer program," he says. "I mean, I was twenty-sixth in the nation in the eighth grade, and you have to be in the top twenty-four among all high school students to come to the training camp. So it was sort of like a dream for me."

In high school he took the American High School Mathematics Examination, which has since been replaced by the AMC 10 and 12 tests. He did well as a freshman, but the breakthrough came during his sophomore year, when he qualified for the summer training camp. The next near he finished in the top twelve in the nation on the USAMO and was honored at the State Department dinner, though he did not make the U.S. team. Not until his last year of eligibility, after five years of hard training and a final grueling runoff exam, did he make it to the top.

Given his accomplishments, you might think that Ian would be a tightly wound ball of competitiveness. Yet in person he doesn't seem competitive at all. He is soft-spoken, generous, close to his parents, a reflective thinker about mathematics and the issues related to math. During the trip on the *Cherry Blossom* to see the fireworks, he was the first to withdraw from the intensely played games. He lay on the floor, put his head in his mother's lap, and fell sound asleep. "I don't think of him as a very competitive person," Ian's father says. "He gets interested in something, and he always tries to do well in whatever he's interested in."

Ian says he tries not to let his competitiveness get the better of him. "After the Mathcounts competition, my teacher said, 'Don't worry about it, Ian. If there's one thing you need to learn, it's to never "what if" yourself.' That's really stuck with me. After a few weeks my disappointment with Mathcounts was behind me. My teacher knew me. She knew that I can often be too hard on myself."

▽

Psychologists have developed many elaborate explanations for human motivation. We are motivated by the desire to demonstrate our autonomy, bolster our self-confidence, or connect with other people. We expend extra effort because we do something well or not well enough. We are intrinsically motivated by interest in an activity or extrinsically motivated by a reward, like praise or a trophy. We are born with genetically determined levels of motivation, or we become motivated through formative experiences.

None of these theories seemed to explain the Olympians' behavior particularly well — or, rather, they all did. The members of the team were so driven to do mathematics that almost any motivating force could be elicited through persistent questioning. In that regard, their competitiveness was unremarkable — as much a part of them as the clothes they wore. One of the favorite activities of the Olympians at the training camp — and later at the Olympiad itself — was to gather around a computer and play an Internet game called Word Racer. An array of random letters is displayed on the screen, and the challenge is to connect adjacent letters to form words faster than your on-line competitors. Usually four to eight players scattered around the world are playing each game, and because the game is highly addictive, some of the players are extremely good. The instant the letters appear, the list of found words begins scrolling frantically down the screen. The Olympians put one person at the keyboard while the others gathered around and called out words. They usually won, but not always — and they hated to lose.

Despite its abstraction, mathematics can be a surprisingly competitive endeavor. Even the casual afterschool math clubs in many middle and high schools can turn into intellectual battlegrounds, depending on how the club is run. Kids who don't think of themselves as good at anything other than math suddenly have a chance to prove their mettle with pencil and paper. Students earn spots on their school team by beating other students, and then those teams compete against other teams. Competitions such as Mathcounts and the AMC tests further the sorting process, with only the best problem solvers moving on to the next level. Even the six kids chosen for the U.S. Olympiad team are scored individually at the end of the competition. All those who don't make the team, a group that includes some of the best high school mathematicians in the country, may end up feeling like losers.

The hard-driving nature of math competitions can be especially distasteful to girls. Even if they like math, they may become disenchanted by math clubs that emphasize speed, rankings, and victory above all. Some observers have suggested that the dearth of girls at the highest levels of mathematical achievement has nothing to do with their math skills but results simply from their disinclination to butt heads with the boys.

The great guru of anticompetitiveness in the United States is Alfie Kohn, an author and lecturer who lives in Cambridge, Massachusetts. For more than two decades Kohn has been on a crusade to eliminate competition from American life. He begins his book *No Contest: The Case Against Competition* this way: "Life for us has become an endless succession of contests. From the moment the alarm clock rings until sleep overtakes us again, from the time we are toddlers until the day we die, we are busy struggling to outdo others." For Kohn competition is a "disease," an "obsession," "warlike," and "destructive." From a very early age, children are taught by parents and other adults to struggle and overcome. Kohn writes,

> Competitiveness is particularly pervasive in our schools, where it is used to prime young students for the rigors of capitalistic struggle. . . . Few values are more persistently promoted in American classrooms than the desirability of trying to beat other people. Sometimes this lesson is presented with all the subtlety of a fist in the face, as with the use of spelling bees, grades on a curve (a version of artificial scarcity in which my chance of receiving an A is reduced by your getting one), awards assemblies, and other practices that redefine the majority of children as losers.

Kohn bases his critique of competition on four contentions. The first is that competition is far from an inherent and necessary part of human existence. On the contrary, Kohn says, we learn to be competitive because society teaches us to be so. Some human groups, such as some of the nonindustrialized cultures studied by anthropologists, seem to be quite free of competition. But in western societies, and especially in the United States, competition has become a religion. Popular culture worships the winner, from the countless movies celebrating the unlikely victory of an underdog to the steady stream of books about getting to the top. Jobs, recreational activities, college admissions, and even parental affection are all structured as contests ("let's see who can get his room cleaned up fastest"). In this way the urge to compete becomes so deeply ingrained in our psyches that it seems to have been placed there before we were born.

Kohn's second claim is that competition does not necessarily lead to improved performance. In fact, competing often makes people perform less well than they would have otherwise. In one study, college students who were trying to develop creative solutions to problems did better when they were not competing than when they were trying to outdo each other. A study of scientists found that the most competitive people tended to be the least productive (as judged by the admittedly problematic indicator of how often their papers were cited by others). Grade school children competing with each other to produce artwork made significantly less complex and creative collages than did children who were not competing. Many other studies have shown that the anxiety and sense of dread that many people feel during contests tends to interfere with their concentration and creativity.

These negative emotions buttress Kohn's third argument, which is that most of the time competing isn't even fun. Turning a diversion into a competitive sport converts something that can be enjoyed for its own sake into a grim, unrelenting struggle, he says. Little League games, according to Kohn, are "institutionalized child abuse." Competition can contribute to ulcers, suicide, and drug abuse. Even for the winners in a competition, the thrill associated with victory may be suspect. "The pure pleasure of competitive triumph is first cousin to the pleasure of punching someone," he writes.

Finally, Kohn disputes the idea that competition builds character. A single-minded focus on winning undermines our sense of fairness and generates a hard-to-resist temptation to cheat. More broadly, we compete because we are trying to prove to ourselves that we are worthy, but we can never overcome doubts about our own capabilities, Kohn says, because there will always be someone who is better. And when we do lose, as inevitably we must, our sense of self can come unraveled: "Competitive loss is a particularly noxious kind of failure, one that contains messages of relative inferiority and that typically exposes one to public judgment and shame."

Kohn's arguments are obviously exaggerated, but any fair-minded observer must concede that he has a point. Competition often assumes ridiculous proportions in our society. Parents dress five-year-old girls in provocative dresses and eyeshadow so they can compete in beauty contests. Parents, teachers, and coaches pit school-age children against each other in head-to-head combat, believing that somehow the experience will make everyone want to excel. In the battle to get into good colleges, high school friends can become rivals who hope that everyone else bombs on their SATs. And competition has an unfortunate tendency to crowd out cooperation. In the past few years the organizers of the Ultimate Frisbee world championships have even had to station "observers" around the field to settle disputes that the players aren't able to resolve themselves.

Kohn's preferred solution is to eliminate competition from our lives. He won't brook the suggestion that competition and cooperation could be more appropriately balanced. In his view, achieving a goal at someone else's expense is ultimately destructive: "The phrase *healthy competition* is actually a contradiction in terms," he writes. "I believe the case against competition is so compelling that parenthetical qualifications to the effect that competing can sometimes be constructive would be incongruous and unwarranted."

One response to Kohn's arguments is to dismiss them simply as utopian and unworkable. If anything, modern life is becoming more competitive, not less. In capitalist and democratic societies, people rely on competition to help them make decisions about products, ideas, and candidates, and as capitalism spreads around the world, so does the intensified competition associated with open markets. Perhaps hunter-gatherer societies were noncompetitive (though many anthropologists dispute this point), but most people are glad that they no longer have to hunt and gather to survive. If the benefits of modernity require some level of competition, most people are happy to make the trade-off. And in a world of continued scarcity and injustice, how should economic goods be distributed if not through some form of fair competition?

But in the end an argument based on the inevitability of competition is unsatisfying. The case for competition can stand only if it can be shown not to be an unmitigated evil.

The first observation that must be made is that many people like to compete. They say that it inspires them to work harder, reach for higher goals, or satisfy a deep longing for recognition. Some choose the direct confrontations of chess or tennis; others prefer the gentler striving of Ultimate Frisbee. Eliminating competition from the lives of such people would not make them happier. Maybe people who like to compete are simply brainwashed victims of our hypercompetitive society. But they don't see themselves as victims and would likely take offense at being labeled insecure neurotics.

Occasionally the well-meaning organizers of math competitions take steps to reduce the competitiveness engendered by individual rankings — say, by having competitors solve problems as a team. But if they go too far in that direction, they quickly meet resistance. Kids like to know where they stand, and they instantly know if they are being patronized. Mathematical reasoning ultimately takes place in the heads of individuals, not in some amorphous group-mind. Competition acknowledges the uniqueness of each person's efforts, and many people are reluctant to give that up.

Furthermore, most people would disagree with the contention that anything less than absolute victory means failure — otherwise, why would sixteen thousand people run in the Boston Marathon each year? Math competitions offer many rewards in addition to being number one. People learn to fit themselves into competitive hierarchies and not to be unduly tormented by their ranking. Indeed, one argument for competition is that students need to learn this skill (though in our society they already get plenty of practice at it).

Competition also has an important social function. Relatively few competitions take place unobserved. People watch the conflicts going on around them and draw lessons from what they see — whether of the nobility or of the futility of ambition. Watching others compete teaches us both what is permissible and what is possible. Competitions let us speculate about our capabilities and our limitations.

In that way competition is both a destructive and a creative force. It tears down the status quo, however inconsequential and temporary that status quo may be, and replaces it with something new. Many highly creative people are consumed by competitive fervor. Sometimes that energy tears them apart, but in other cases it produces works of great art, literature, science, or mathematics.

Competitions can be structured in ways that pose less risk to the self-esteem of competitors. Math contests don't always have to compare one student to another. Including a team component can reduce the anxiety of head-to-head comparisons, just as Word Racer is a different game when played against a roomful of kids in Bombay than when you're competing against your best friend at the next computer terminal. Or everyone can strive to reach a high but attainable standard — as when a single problem is posed and students have as long as they need to solve it. By focusing on the task rather than on victory, competitions can help students feel good about what they can do rather than about whom they can beat.

These observations apply even at the very highest levels of competition. When Melanie Wood began attending the Olympiad summer training camps, she was shocked by the competitiveness she found. "I've always had a sort of anticompetitive position," she says. "At the training camp there are these team contests where students work in teams against other teams. I thought that was too competitive — and in fact for a couple of years those competitions kind of dissolved, which was blamed on me. I didn't really have anything to do with it, but it's true that the anticompetitiveness that I tried to instill in other people was partly responsible for their dissolving.

"I've always been opposed to summing up scores and saying that this is who wins and this is who loses. I like to work with people; I don't like to wonder if I can beat them. I mean, I don't mind saying, 'Oh, we're going to beat the Bulgarians or the Russians,' because I don't know them. But I have a lot of issues with not working together with someone else on your team.

"Some people say that I'm very competitive — and that's true — but there's an important distinction. I like to do very well, but I like other people to do well, too. It's never been for me about beating other people."

▽

The second problem on the Forty-second Olympiad was an inequality, a mathematical expression in which one quantity is greater than or less than another quantity. It referred to three positive numbers labeled a, b, and c. For any value of those three numbers, from zero to infinity, the contestants had to prove that

$$a/\sqrt{a^2 + 8bc} + b/\sqrt{b^2 + 8ac} + c/\sqrt{c^2 + 8ab} \geq 1.$$

(The symbol $\geq$ means "greater than or equal to.")

   This problem is not as hard to understand as it looks. Say that a equals 1, b equals 2, and c equals 3. In that case, the first term in the problem is equal to

$$1/\sqrt{1^2 + 8 \times 2 \times 3}, \text{ or } 1/\sqrt{1 + 48}, \text{ which equals } 1/\sqrt{49}, \text{ or } 1/7.$$

The second term is equal to

$$2/\sqrt{2^2 + 8 \times 1 \times 3}, \text{ or } 2/\sqrt{28},$$

and the third term is

$$3/\sqrt{25}, \text{ or } 3/5.$$

The sum of those three fractions is about 1.12, so the inequality holds for those three numbers. But the Olympians needed to prove that it holds for an infinity of different numbers, not just a particular three.

   Ian began his assault by inserting some random numbers into the problem. He thought he might be able to learn something about the inequality by seeing how close he could get the two sides to being equal. But that approach offered no immediate clues about how to begin constructing a proof. He then tried rewriting the inequality to eliminate the square-root signs but immediately ran into difficulties. The problem can be solved that way, but it's not easy. One of the Chinese team members used a similar approach to raise one number to the 480th power and then take the 486th root of the result. The method works in the end, but it's an ugly solution with many potential pitfalls, and Ian deemed the approach too risky.

He began to worry. Though he had solved the first problem pretty quickly, the third problem looked so simple that he knew it would in fact be very hard. To leave enough time to finish the exam, he needed to make some progress. "I thought it shouldn't be that tough," he later said. "There are only a limited number of things you can try. The standard way would be to multiply the terms and square them to get rid of the square-root signs. But I could see that if you did that you'd end up with a mess. So I knew I had to come up with something clever."

About this time some of the Olympians started to take bathroom breaks. Going to the restroom was complicated, since the competitors were not allowed to consult any cheat sheets they might have sneaked into the exam. They had to raise a hand to attract the attention of one of the "invigilators" who prowled among the tables like sharks among a school of fish. Then a monitor accompanied each one to the bathroom and back.

Ian was returning from the bathroom when he thought of it. "Jensen's," he said to himself. And he knew right away that the problem was solved.

Johan Jensen was the head of the technical department of the Copenhagen Telephone Company from 1890 until the year before his death in 1925. The son of an educated but feckless father, he was essentially self-taught in mathematics and never held an academic position. But at the College of Technology in Copenhagen, where he studied science, he fell in love with mathematics and decided to devote his nonprofessional life to it.

In 1906 Jensen published an article that guaranteed his mathematical immortality. He showed that for a given class of mathematical relationships, a particular inequality (described in the appendix) holds. What Ian saw in his moment of illumination was an extremely clever way of adapting Jensen's inequality to problem two. He took the pieces of the problem and plugged them into Jensen's inequality. After performing a series of algebraic manipulations, he derived something close to the following expression:

$$\frac{a}{\sqrt{a^2+8bc}}+\frac{b}{\sqrt{b^2+8ac}}+\frac{c}{\sqrt{c^2+8ab}}\geq\frac{\sqrt{(a+b+c)^3}}{\sqrt{a^3+b^3+c^3+24abc}}$$

(Actually, Ian used something called cyclic notation, which compresses this expression a bit, but the result was essentially the same.)

At this point the problem was almost solved, because the term on the left-hand side of this inequality was the same as the term in the original problem. If Ian could prove that the right-hand side was greater than 1, he was done. That meant proving that the numerator of the fraction, $(a + b + c)^3$, was larger than the denominator, $a^3 + b^3 + c^3 + 24abc$. That was not so hard to demonstrate. Ian proved it using something called the arithmetic mean–geometric mean inequality. He did a few straightforward calculations and the proof was complete. "As desired," he wrote at the end of his solution.

As with Tiankai's solution to problem one, the proof seems straightforward. But the judges admired Ian's solution even more than they had Tiankai's. Before an Olympiad the judges typically prepare a list of different solutions for each problem to help them decide whether a student deserves full credit for a problem, partial credit, or none at all. When the judges received Ian's solution at the end of the first day, they saw that he had developed a completely new way of solving problem two. None of them had realized that it was possible to use Jensen's inequality as Ian had. And his approach was simpler and more elegant than anything they had concocted: his solution may very well be what Erdős would have called the Book proof for problem two.

When someone performs well under difficult circumstances, we can admire that person for his skill and pluck. When someone performs well under intense competitive pressures, when the eyes of the world are focused on that person's every action, our admiration turns into something closer to awe. And perhaps that's the greatest argument in favor of competition. It can produce moments not only of great achievement but also of great beauty.

5. talent

Of the three problems posed each day of the Olympiad, the first is traditionally the easiest, the second the next hardest, and the third the most difficult. So what were the Olympians to make of problem three?

> Twenty-one girls and twenty-one boys took part in a mathematical competition. It turned out that (a) each contestant solved at most six problems, and (b) for each pair of a girl and a boy, there was at least one problem that was solved by both the girl and the boy. Prove that there was a problem that was solved by at least three girls and at least three boys.

In its self-reflective simplicity, this was the perfect Olympiad problem. Yet its inclusion on the exam had been a matter of great dispute. The six problems are chosen through a complicated and highly political process. The coach of each team comes to the Olympiad site a few days before the event with a list of possible problems. As soon as they have checked into their hotel rooms, the coaches begin a series of closed-door meetings to decide on the problems that will appear on the test. (After the choices are made, the coaches are quarantined from the competitors until after the exam and the assistant coaches lead the teams.) During these conclaves many of the coaches from weaker teams lobby forcefully for easier problems. If their team scores poorly, it reflects badly on them and on their country. Though every problem on the exam has to be of Olympiad caliber, the odds of embarrassments are lower if the problems are easier rather than harder.

What would become problem three on the Forty-second Olympiad was one of the most simply stated problems in the event's history. But the assembled coaches quickly realized that it would be extremely difficult to solve. Furthermore, it had all the trappings of an all-or-nothing problem. Unless the Olympians solved the problem completely, they probably weren't going to make enough progress to earn partial credit for their solutions, so most of the scores were going to be 7s or 0s.

Titu was one of the coaches who argued passionately for including the problem. It was simply too beautiful to forgo. "It's a problem that could be understood by any student," he said. "But to solve it, that's a different story."

▽

Everyone on an Olympiad team is mathematically "talented." But what is the nature of that talent, and how do Olympians acquire it? Were they born with it, or did they develop it through assiduous practice? Did their talent inevitably reveal itself at an early age, or did it become apparent only later? Were teachers and coaches essential to foster that talent, or would their skills have appeared anyway? The word "talent" is plagued by almost as many ambiguities as the word "genius."

One view of talent is that it is a God-given gift poured into a person before birth. According to this view, such a gift often reveals itself at an early age because it arises from a person's genes, not experiences. In her book *Gifted Children: Myths and Realities,* Ellen Winner, a psychologist at Boston University, describes several striking cases of young children who developed intellectually at a prodigious pace. One infant boy, whom she calls David, could understand questions like "Where's Daddy?" and "Can you go and get it?" when he was eight months old — about a year ahead of the normal developmental timetable. At fifteen months, when most children are just learning to speak, David had a vocabulary of about two hundred words. He taught himself to read when he was three by having his mother sound out words for him as she and he pointed to the words in books, and a few weeks later he was reading on his own for hours on end. By the age of four he was reading biographies and science books and poring over atlases, and he began to write stories and letters to friends and family members. He knew how to add and subtract by that age and became fascinated with the idea of infinity. The next year he gained a deep understanding of fractions and ratios — concepts normally not emphasized until middle school. He also began to learn other languages when he was five — he studied French, Spanish, and sign language in books from the library. By the time David entered first grade he was reading at the sixth-grade level and was far ahead of his classmates in most other subjects.

Winner refers to David as a "globally gifted child." He advanced very quickly in many different areas — math, reading, the sciences, languages. Other fast learners excel in more restricted domains. Some become fascinated by numbers and patterns. They mathematicize their world, as Tiankai did when he began to interpret objects around him as geometric shapes. They factor the numbers they see on license plates or memorize pi to hundreds of places. Other children have an incessant need to draw. They may spend hour after hour sketching trains, ballerinas, or animals. Some musical prodigies, though certainly not all, fall so deeply in love with their instruments that they work their fingers raw; Leonard Bernstein's parents had to plead with him to stop practicing the piano.

Each such child has a different developmental trajectory, says Winner, but they share several common features. All have what she calls an overwhelming "rage to master." When they become interested in something, they devote themselves to it with

every ounce of their being. Many become obsessed with a particular subject — volcanoes, insects, prime numbers, Greek myths
— and develop an insatiable need to learn everything they possibly can about that subject. Some seem to need relatively little
sleep, and they can exhaust their parents with their demands.
They have an enviable ability to achieve the mental state that
psychologist Mihaly Csikszentmihalyi at Claremont Graduate
University calls "flow"; when they are working in their area of
interest they seem to lose touch with the outside world and become one, mentally and physically, with what they are doing.

Even with less precocious children, the view that talent is innate is widely held. Parents see their children as good at some
things and bad at others, especially when comparing siblings.
They can't think of any obvious force driving a child toward a
particular set of competencies. They therefore assume that talents are a reflection of who that child is, as predetermined as the
color of his or her hair.

▽

Some psychologists have always been uncomfortable with this
view. Ascribing high achievement to some mysterious biological
essence makes about as much sense as the Romans' belief in *genius,* they say. People are a product of their experiences, and the
only way to explain talent is to examine those experiences.

One such skeptic was Michael Howe, who was a professor
of psychology at the University of Exeter in England until his
death a few months after the Forty-second Olympiad. In a long
series of articles and books, Howe argued vigorously against
what he called the "talent account": the idea that just a few people are born with a special mental capacity that enables them to
achieve high levels of performance in a particular field. One feature of the talent account, according to Howe, is the belief that
coaches, teachers, and other adults can detect these special qualities when a child is young, even before he or she has demon-

strated exceptional levels of performance. These talented young people can then be given special attention to develop their genetically based gifts. Those who aren't born with talent can develop their skills, but they can't expect to keep up with their more talented peers.

The evidence simply does not support the talent account, Howe insisted. Individuals are good at certain tasks because of previous experiences, not because they have a special talent or gift. An unusual achievement is like the tip of an iceberg — people see the part that is plainly visible but overlook the huge base of practice and thought that support it. Howe and several of his colleagues were especially interested in music, and they pointed out that all students require about the same amount of practice time to reach particular musical milestones (as measured, for example, by a series of exams). One study of German student violinists showed that individuals training to be concert soloists had practiced for about ten thousand hours by the age of twenty-one (which works out to about twenty hours a week for ten years). The violinists who intended to be teachers rather than performers had practiced for about half that time, but no one had become an accomplished violinist without engaging in many hours of rigorous training.

Studies of other fields have produced similar results. It takes about ten years of dedicated practice to become a high-level chess player, ballerina, physicist, or mathematician; even the youngest member of the U.S. Olympiad team, Tiankai Liu, had been thinking deeply about mathematics for more than ten years. In a study of seventy-six major composers, all but three had spent at least ten years composing before they began to produce major works, and the three exceptions (Dimitri Shostakovich, Niccolò Paganini, and Erik Satie) took nine years. "Although it is widely believed that certain gifted individuals can excel without doing the lengthy practice that ordinary people have to engage in, the evidence contradicts that view," Howe concluded.

Picking out young people who will go on to excel is also much tougher than parents, teachers, and coaches think, Howe pointed out. In the 1980s, for example, Lauren Sosniak interviewed twenty-one outstanding American pianists in their mid-thirties and their parents. She found that few of the future musicians had displayed signs of exceptional promise when they were young. Even after they had been playing the piano for six years, confident predictions about their eventual eminence would have been possible in only a few cases.

When adults say they see talent in a child, they may actually be seeing a child who has matured earlier than others, either physically or mentally, Howe and other psychologists have said. Or the child may simply find a task more interesting or may have had some previous exposure to it. Whatever the reason, as soon as the child is labeled as talented, the prediction becomes self-fulfilling. That child is given extra attention, so he or she draws farther ahead of the pack.

In fact, most people can achieve at very high levels in a particular domain if they are willing to invest the necessary time and energy, say psychologists who share Howe's perspective. For example, Anders Ericsson at Florida State University cites studies in which randomly selected people learned to do a particular task so well that observers assumed they had a special talent for it. With the proper training, people could memorize numbers and words, perform calculations, or tap out patterns with astonishing skill. They didn't need an inborn talent; they simply had to engage in deliberate, purposeful, and carefully monitored practice.

What about children who demonstrate unusual abilities at a very early age, such as talking before the age of one or learning to read before the age of four? Howe expressed doubts about many of these reports. Such recollections almost always come from parents or from grown-up prodigies, which inevitably raises questions of veracity. Howe also was skeptical that these prodigious abilities really developed without prodding. Prodigies almost always emerge from families that are particularly receptive to the idea of raising a prodigy. The parents provide bountiful stimulation to their children — books, music, lessons, conversation, playthings. They focus great energy on their children and expect them to do well. Sometimes the parents change jobs, move to different cities, or give over a substantial portion of their own lives to developing their children's gifts. At a symposium in 1993 Howe said, "There are parents who take the credit for their child's achievement, but there are others who say that the child received a gift from God and that they had nothing to do with it at all. As proof of this, they will say that on one day their child had 623 words in his vocabulary and five days later he had 678, but that they knew nothing about why the child did so well."

The bottom line, Howe and other psychologists have concluded, is that people do not have inborn gifts or talents that predestine them to greatness. Those who are highly accomplished shape their lives with the same basic materials that the rest of us have. "The notion of innate talents may turn out to be entirely superfluous," Howe wrote. "Even though the idea that innate talents provide a mechanism via which genetic differences between people have impacts on their capabilities is widely accepted and commonly believed in, there are good reasons for thinking that such talents are mythical rather than real."

In his book *Genius Explained,* Howe subjected his ideas to the acid test: accounting for the achievements of the young Wolfgang Amadeus Mozart. Born in Salzburg in 1756, the youngest son of an ambitious and oppressive father and a doting and oppressed mother, Mozart embodies more than any other composer the popular understanding of genius. Under the watchful eye of his father, Leopold, who was an accomplished violinist, composer, and assistant conductor in the employ of the prince-archbishop of Salzburg, Mozart learned to play the harpsichord when he was three after watching his older sister, Marianne, practice. Leopold gave his young son a notebook of easy tunes that he had compiled for Marianne, and Mozart quickly mastered the pieces. "Wolfgangerl learned this minuet and trio one day before his fifth birthday in half an hour at half past nine in the evening of January 26, 1761," his father jotted down in the notebook. Not long thereafter, Mozart produced what may have been his first compositions: two short pieces for clavier that his father also copied into the notebook.

Between the ages of seven and almost eleven, Mozart and Marianne traveled with their father through Europe, demonstrating their virtuosity on the harpsichord and violin. Mozart would astound and delight his audiences by sightreading compositions he had never seen before, or adding a bass part and intermediate voices to a melody he was given, or playing the harpsichord with a cloth covering the keys. He could improvise on a theme for hours, and he became a fluent composer, dedicating his youthful sonatas, symphonies, and concertos to the queens, lords, and princesses who marveled at his exploits. "Nowadays, people ridicule everything that is called a miracle," Leopold Mozart wrote in a letter to a friend. "Hence one has to persuade them; and it was a great pleasure and a great victory for me to hear a Voltairian say to me, 'Now for once in my life I have seen a miracle; this is the first.'"

Mozart also had an incredible memory for music. At fourteen he wrote out the complete score of a long multipart musical composition, Gregorio Allegri's *Miserere,* after hearing it played on just a few occasions. He could compose and improvise in the style of any music he heard, and he could mimic the sounds of the birds or people around him. He was absorbing the musical environment of his age, preparing to become, as an adult, the author of some of the most beautiful music ever written.

Howe did not dispute that Mozart had begun writing music

at a very early age. But, he pointed out, those early works were not outstanding by the standards of mature composers, though they gave some glimmers of what was to come. Also, those early works were written down by his father, who might have improved them in the process. Many of his childhood compositions draw heavily on the work of other composers. Mozart's earliest symphonies, for example, are strikingly similar to those of Johann Christian Bach, who encouraged the eight-year-old Mozart when they met in London in 1764. None of Mozart's major works appeared until he had been composing for more than a decade. For example, his earliest concerto that is considered a masterpiece, the Piano Concerto No. 9 in E-flat major (K. 271), was composed when he was twenty-one.

Similarly, Mozart's performing abilities were surely extraordinary, but they weren't inexplicable, Howe said. No one knows how much the young Mozart practiced, but his father imposed a strict regimen on both Wolfgang and Marianne, since the two of them had become the main source of income for the household. Mozart had essentially no friends as a child and devoted most of his time to music. Howe assumed, probably conservatively, that he practiced about three hours a day from the age of three. In that case, by the time he was six, when he went on his first musical tour of Europe, he would have had 3,500 hours of practice. That's about how long it takes for a young performer to become a very good amateur.

That amount of practice time would have been very unusual for a child of the eighteenth century. But since then it has become more common. Today, Howe claimed, many children reach the same levels of performance as did the young Mozart. His earliest feats were certainly impressive, but they would not attract nearly as much attention if they were repeated today.

Finally, although Mozart's ability to absorb and retain music was certainly extraordinary, Howe pointed to a considerable body of research showing that people can remember huge quantities of information if they are knowledgeable in that area. Chess experts, for example, can recall vast stores of moves and strategies from past games that they and others have played. And everyone can develop, through training, much better memories for particular kinds of information.

Mozart's feat with Allegri's *Miserere* was remarkable, Howe wrote in his book *Genius Explained,*

> but imagine the unusual everyday life of the young Wolfgang Amadeus Mozart. He inhabited a world of music, hour after hour, day after day, in the company of a father who was an expert teacher. By adolescence, the sheer amount of Mozart's musical knowledge would have been enormous by most people's standards. He would have recognised many familiar structures and patterns, eliminating the need to recall each note separately. As a result, compared with a non-musician Mozart would have perceived the task very differently, with the information that needed to be remembered being meaningful and interconnected. And although Allegri's *Miserere* is a lengthy composition, it is one that happens to contain a great deal of repetition. For a person as knowledgeable as Mozart, that would have lightened the burden of remembering.

People obviously have inborn differences that are the product of their genes, Howe said. But the genetically based differences that contribute to achievement may not be the ones most people usually cite. Rather than having a music gene or a tennis gene or a math gene, some people may be born with the ability to concentrate more intently and not be distracted. Some children may be more optimistic and self-confident, allowing them more easily to overcome the obstacles facing anyone who tries to do something extremely well. They may have higher levels of energy and enthusiasm, so that they dedicate themselves to tasks with greater purpose. In other words, the differences that influence achievement may be related to how people feel rather than how they think. As Howe wrote, "A number of geniuses, including Darwin and Einstein, have disclaimed having superior inherent intelligence, but no genius has ever denied either possessing or relying upon a capacity for diligence or a healthy curiosity."

These essentially emotional factors may be reinforced by feedback loops similar to those involved in the selection of talent. Imagine, for example, that one infant's brain experiences music as particularly soothing or interesting. Such a child might be more attentive to music from a very early age (even in the womb), and the parts of the brain specialized for musical abilities would therefore become more highly developed. Or a child might have an aesthetic preference for shapes and patterns, which in turn would serve as the seed of future mathematical ability. There may be critical periods during a young child's life, perhaps lasting just a few days or weeks, when experiences are indelibly engraved on the developing personality. All of these possibilities seem just as plausible as the idea that people are born naturally good at something.

Upon these internal processes must be superimposed all the unique occurrences of an individual's life. The direction provided by parents, teachers, and coaches obviously has a huge influence. In one study, students were randomly assigned either to be tutored individually or to be taught in a conventional classroom. The average tutored student performed better than 98 percent of the students taught the conventional way — two standard deviations above the classroom norm. Many of the Olympians point to a specific person — a teacher, parent, mentor, or coach — who provided critical feedback and guidance. In a young person with ambition and determination, such an adult can create talent where none was evident before.

And then there is sheer luck, which influences everyone to a greater or lesser degree. The childhood illnesses of Charles Dickens isolated him from other young people and put him in the company of books. If Charles Darwin had not volunteered to accompany Captain Robert Fitzroy on the voyage of the *Beagle*, Alfred Russel Wallace would probably have been the discoverer of the theory of evolution. Every Olympian recalls critical moments that had immense consequences. A coach needs one more member to round out a math team and asks a seventh-grade girl who he has heard is good at math. A group of mathematicians starts a lecture series that attracts the interest of a particular boy and his parents.

∇

The basic contention made by Michael Howe and like-minded psychologists — that the achievements we ascribe to talent arise from practice and experience, not from inborn abilities — has always generated great opposition. Life just doesn't seem to work that way. Some kids learn how to draw, play basketball, or multiply more easily than others. Maybe the difference is subtly motivational rather than cognitive, but the end result is the same: some children have to work very hard to master a new skill while others seem to pick it up effortlessly. As psychologists David Feldman and Tamar Katzir of Tufts University once wrote in a commentary on one of Howe's articles: "If anyone can prove that the works of these individuals can be explained without recourse to a construct like natural talent, we will concede that talent does not exist: Mozart, Picasso, Shakespeare, Martina Hingis, Baryshnikov, Pavarotti, Ramanujan, Judit Polgar, Michael Jordan, and Robin Williams. Practice, indeed."

The practical implications of rejecting the talent account also bother some critics of Howe's work. Resources to develop talent are inevitably limited. Focusing resources on children who show interest and ability in a given domain would seem to make more sense than trying to spread those resources among all young people. And the idea of inborn talent can serve a valuable social end, even if it seems discriminatory. A young person in disadvantaged circumstances may feel so strongly that talent will prevail that he or she will work ceaselessly to beat the odds.

Other critics take issue with Howe's use of the data. According to Ellen Winner, reports of early achievement among infants and young children are too numerous to ignore. "These kids are incredibly motivated to learn," she says. "The average child learns to read between six and seven years of age with considerable instruction, but some children learn to read at age three or four with minimal instruction. They've broken the code, something clicks. You can't bribe an ordinary child to do that."

Evidence also continues to accumulate that the brains of at least some highly able people are different from other brains, Winner observes. The observation that a disproportionate number of mathematicians, musicians, and artists are left-handed or ambidextrous suggests that their brains are organized differently. People who are not strictly right-handed tend to have more activity in the brain's right hemisphere, which is typically specialized for spatial rather than verbal representations. Maybe the brains of these individuals are predisposed to carrying out particular tasks.

Finally, the critics counter Howe's arguments by pointing to what is called savant syndrome. Savants have brain dysfunctions that cause severe social or intellectual impairments; they typically have IQs ranging from about 40 to 70. But they are exceptionally skilled in particular areas. One autistic savant, a man named Richard Wawro, who lives in Scotland, produces paintings acclaimed as modern masterpieces. Another, Kim Peek, who was the model for the character played by Dustin Hoffman in the 1988 movie *Rain Man,* has memorized parts of more than 7,600 books, but he is so developmentally disabled that he must rely on ther for his daily needs. An intellectually disabled savant named Alonzo Clemons can produce a perfect sculpture of an animal that he sees just for an instant. Such abilities would be remarkable in an otherwise ordinary person. In individuals with the deficiencies characteristic of savants, they seem miraculous.

One of the most famous savants in American history was a slave named Thomas, who was born in 1849. When Tom was very young, his parents were purchased by General James Bethune of Columbus, Georgia, and by tradition Tom was given his owner's last name. Blind and intellectually impaired from birth, Tom grew up in the Bethune household listening to the sounds around him, including the singing and piano practicing of the general's daughters. One day when Tom was four years old, he sat down at the piano and began to imitate perfectly the chords he had heard the girls practicing. General Bethune arranged for Tom to receive music lessons, and by the age of six he was improvising and writing original musical compositions. In 1857 General Bethune rented a large hall in Columbus, and Blind Tom, as he came to be called, began performing before awestruck audiences of southern aristocrats.

At the age of nine Tom was hired out to a promoter, who sent him to hundreds of cities across the United States, performing four shows a day at each stop. His abilities were legendary. Musicians in the audience were invited to perform a piece of music, after which Tom would play the same piece. He could play a song and then turn his back to the piano and play the same piece with his hands reversed. Tom could sit next to someone playing a treble part and improvise a bass accompaniment as the other person played. If someone struck ten different keys on the piano simultaneously, Tom could instantly list all ten notes. During his life he memorized thousands of songs, and some of his compositions are still performed today. Newspapers called him the greatest pianist of his time, with skills surpassing even Mozart's.

After the Civil War Tom was free, and General Bethune became his manager. He provided Tom with food, shelter, musical
instruction, and an allowance of twenty dollars per month
(Tom's performances brought in an estimated eighteen thousand
dollars each year). When Tom turned twenty-one, the general arranged to have him declared insane and became his legal guard-
ian. For the rest of his life, until his death at the age of fifty-nine
in Hoboken, New Jersey, Tom traveled widely, first with General
Bethune and later with one of the general's daughters-in-law, giving demonstrations of his prodigious abilities. In photographs he
is a short, heavyset man with close-cropped hair, closed eyes, and
the expression of someone listening to a phrase of music playing
inside his head.

In 1869 Mark Twain saw one of Tom's performances and
described it this way:

> He swept [the emotions of his audience] like a storm, with
> his battle-pieces; he lulled them to rest again with melodies
> as tender as those we hear in dreams; he gladdened them
> with others that rippled through the charmed air as happily
> and cheerily as the riot the linnets make in California woods;
> and now and then he threw in queer imitations of the tuning
> of discordant harps and fiddles, and the groaning and wheez-
> ing of bag-pipes, that sent the rapt silence into tempests of
> laughter. And every time the audience applauded when a
> piece was finished, this happy innocent joined in and clapped
> his hands, too.

Taken together, the arguments against Howe's position seem
very strong. Yet Howe and his supporters have counterargu-
ments to every criticism. If talent is so obvious in the young, why
do so many precocious children fade from the scene while high
achievers emerge from nowhere? Many exceptional performers
show no signs of brain abnormalities, so an unusual brain does

not seem necessary for excellence. Even the feats of savants do not necessarily imply that talents are innate, Howe said. Such individuals are not born with the skills they master. Like anyone else, they must learn how to play the piano, paint, or recite long prose passages. As a very young child, Blind Tom was fascinated by sounds. For hour after hour, in the shuttered isolation of his mind, he listened to the singing of General Bethune's daughters, their scales and chords on the piano, his mother's voice. When he was finally allowed to sit at the piano and play, he must have felt that he was being given an opportunity to speak after years of silence.

▽

So the old argument between nature and nurture remains as inconclusive as ever. You could look at the members of an Olympiad team and say, "These kids obviously were born with a gift that enables them to solve these problems." Or you could look at them and say, "Anyone who spent as much time thinking about mathematics as these kids do could get to this level." The question seems insoluble: are our genes or our experiences primarily responsible for our accomplishments?

According to David Moore, a psychologist at Pitzer College and Claremont Graduate University in California, this question has never been answered, despite decades of effort devoted to it, for a good reason: the question makes no sense. We cannot ascertain the extent to which genes contribute to a complex human trait like intelligence or creativity. The question is not a scientific one; it is pseudoscience.

Moore acknowledges, of course, that no one sees complex traits as originating entirely in genes or entirely in experiences. Even the most ardent behavioral geneticists concede that experiences deeply shape our lives. But Moore takes the argument much further. He says that the nature of the interactions between our genes and our experiences sharply limits the kinds of questions we can ask and the kinds of answers we can expect. Distinguishing between biological and cultural influences on human traits is the wrong way to think about the world, he says. "Psychology is still a very young field, so we are dealing with these very large constructs that need to be broken down before we can understand them."

In his book *The Dependent Gene: The Fallacy of "Nature vs. Nurture,"* Moore points out that people tend to array human traits along a sort of spectrum. At one end we place traits like hair color or upright stance that seem to be 100 percent genetic. At the other end we place traits that seem to have nothing to do with our genes, such as the languages we speak and the clothes we wear. In between are intermediate traits, like intelligence or shyness, that seem to have some genetic and some environmental components. For example, books and magazine articles often claim that intelligence is 50 percent genetic and 50 percent environmental (or 70 percent genetic, or 30 percent, depending on the point the author is trying to make).

But this is the wrong way to think about traits, Moore insists. Of course our genes are involved — how could they not be? But from the moment of conception on, genes function only in response to the environment around them. At first this environment is limited to the fertilized egg, in which various proteins and signaling molecules interact with the genes and cause them to express particular proteins. Later, when the egg begins to divide and attaches itself to the uterine wall, the environment becomes more complicated. Now the developing embryo can be influenced by many different environmental factors: the health of the mother, exposure to viruses or chemicals like alcohol, the foods the mother eats, immune reactions between the mother and embryo, the presence of a twin, even such factors as the season of conception or the sounds an embryo hears. Random events before birth — even the division or movement of a single cell —

can send development down one pathway rather than another. "From the moment of conception," Moore writes, "environmental factors and genetic factors are in an ongoing dialogue with one another about building a person. Each of these sets of factors brings its own necessary information to the conversation."

When a child is born, the complexity of this dialogue increases by orders of magnitude. Now the influence of the environment is ubiquitous, with countless and unpredictable effects. Even in traits that seem to be wholly the product of our genes, closer examination reveals the environment's critical role. Hair color, for instance, is influenced by the amount of copper in the cells that produce hair. If a person's diet includes very little copper, hair color progressively lightens.

How we walk is another example. "When people say there are certain traits that appear to be unaffected by the environment, it's because there are certain characteristics of all human developmental environments that are identical," Moore says. "For instance, it seems like our tendency to walk, our two-legged gait, is genetically programmed, and people who used to study motor development talked about a thing called a central pattern generator in the brain that was responsible for our two-legged gait. But it turns out that when the astronauts got to the moon they discovered that it was really inefficient to try to walk with that gait in a low-gravity environment, so they immediately stopped walking and started the kind of hopping movement that you see in films. Obviously this has never been tested, but I assume that if you raised humans in a low-gravity environment, they would walk differently than we do."

Research with laboratory animals confirms the pervasive influence of the environment on the development of traits. For example, many monkey species are so terrified of snakes that it has been considered an inborn trait. But monkeys that are born and raised in a laboratory and fed a diet consisting only of fruits and monkey chow are not afraid of snakes. Remarkably, though, laboratory-raised monkeys who are given a single live cricket to eat each day, in addition to their fruit and monkey chow, *are* afraid of snakes. Something about the interaction between monkey and cricket triggers the development of the snake phobia.

Or consider the eating behaviors of newborn chicks. When they are just two days old, chicks will approach, pick up, and eat mealworms they see on the ground — a seemingly innate trait if there ever was one. But if the chicks' feet are covered with cloth "shoes" that permit them to walk but not to see their toes, most of the chicks just stare at the worms, one eye at a time. Even with a trait that is critical to survival and would therefore seem to be hard-wired in the brain, the role of the environment is essential.

In fact, all our traits work this way, says Moore. The genetic and environmental influences on a trait cannot be separated. The only logical thing to say is that genes are 100 percent responsible for our traits, and experiences are also 100 percent responsible for our traits. Therefore, scientists cannot hope to parse a trait into an environmental component and a genetic component. The best they can do is to study how genetic and nongenetic factors interact at specific moments to produce a very specific behavior or characteristic.

This view of life, which Moore calls the developmental systems perspective, has important implications for human behavioral research. For example, Moore contends that studying the genetic factors contributing to a broad trait like "mathematical talent" makes no sense because the effects of genes and experiences are far too intertwined to untangle. Therefore, the question "Is mathematical talent inborn or learned?" is not a scientific question. Researchers will have to focus on much more specific skills. "The right thing to do would be to break mathematical competence down into things like visualization, computation, and other factors — who knows what they are, it's so early in the process. Conceivably you could look at the contributions of genes and environments to each factor. There has to be some sort of genetic contribution to these traits — there is to every trait — but I don't think anyone has even come close to figuring out what that might be."

The proper way to think about human traits, Moore says, is to treat nature and nurture as a single system. Our traits emerge as a result of interactions among all of these factors. The information that contributes to the construction of a trait is distributed throughout the system, so no one part of the system can account for a trait. The trait arises from the system as a whole, not from part of it.

▽

Of the 119 individuals who had been on U.S. teams as of the Forty-second Olympiad, no one was as accomplished as Reid Barton. He attended his first training camp the summer after seventh grade. Following his freshman year he was on the Olympiad team that finished third in Taiwan, where he was awarded a gold medal. (The top twelfth — about forty of the five hundred competitors at a typical Olympiad — win gold medals.) The next year, as a sophomore, he won a gold medal in Romania, as he did the following year in Korea. If Reid were to win a gold medal at the Forty-second Olympiad, he would become the first person in the history of the event to win four gold medals.

Reid became interested in mathematics at a very young age. He taught himself calculus when he was nine and scored a perfect 5 on the mathematics advanced placement exam when he was ten. He also was an accomplished pianist and cellist; at the training camp he often passed the time by playing Chopin quickly and loudly. Media accounts of his achievements often describe him as home-schooled, but that's not really accurate. In fact, he's omni-schooled. After elementary school, he began taking a combination of high school and college courses that he, his parents, and an unusually cooperative public school system arranged. Tall, blond, skinny, not an athlete but not at all clumsy (he more than held his own in Ultimate Frisbee games), Reid came across as a friendly, reserved, all-American kid.

Problems one and two took Reid almost two hours, though he was sure of his solutions once he finished them. Then he confronted the puzzling simplicity of problem three. Problems that involve sorting objects into categories fall into a branch of mathematics known as combinatorics. The field originated in the study of probability by Pierre de Fermat, Blaise Pascal, and other seventeenth-century mathematicians, who were inspired not by the beauty of the math but by a very practical concern: they wanted to calculate the odds associated with gambling.

During the eighteenth and nineteenth centuries, combinatorics was something of a mathematical backwater. Its association with gambling accounted in part for its lack of status, but more fundamental forces were also in play. Combinatorics deals with discrete entities, like the cards in a deck or socks in a drawer. But most of the important problems being investigated by mathematicians at that time involved continuous entities, like lines or surfaces. Mathematicians developed many general techniques to analyze such problems, particularly the methods of calculus. They found far fewer general techniques to analyze problems involving discrete entities, because discrete objects can have so many different properties and can be arranged in so many different ways. Each problem has to be broken down into its constituent pieces, from which a useful pattern may or may not emerge.

In the twentieth century combinatorics staged a comeback for one overriding reason: the invention of electronic computers. The objects manipulated by computers are discrete, not continuous. They are the 1s and 0s of on-off switches, not the smooth curves of continuous surfaces. To analyze the capabilities of computers, mathematicians had to develop new ways of dealing with noncontinuous objects, and that need reinvigorated the field of combinatorics.

Reid insists that his experience with computers played no role in his solution to problem three, but in fact he is as proficient at programming as he is at math. During his high school years he had a part-time job in the laboratory of Charles Leiserson, a computer scientist at MIT. At first he worked on Clikchess, one of the top chess-playing programs in the world. When Leiserson took a two-year leave of absence to work at Akamai Technologies in Cambridge, Reid went with him to work on the company's Internet software. Reid has "an excellent sense of aesthetics," Leiserson told writer Dana Mackenzie. "His code is clean, well organized, simple, and easy for other people to modify. It's unusual to find that ability in someone so young."

The trick with many combinatorics problems is to arrange the entities in a way that demonstrates the desired property. But because every problem is different, no general rules exist for deriving such an arrangement. You just have to think clearly and deeply.

Reid looked at the pieces of problem three for a long time. It states that 21 boys and 21 girls have taken a math test, and each of the 42 contestants solved at most six problems, though the total number of problems on the test is not specified. Furthermore, if you took any boy and any girl, you could find a problem that both of them had solved.

Suddenly Reid thought of something. He sketched out part of a table. At the top of each column he listed one of the 21 boys. Then he put the 21 girls into the rows of the table. In each box he put a letter representing one of the problems solved by both that girl and that boy. (The specific letter chosen in this solution is arbitrary, though the appendix contains an interesting story about the choice of solved problems.) The full table contained 21 × 21, or 441, boxes; the upper-left-hand part would look like this:

	Boy 1	Boy 2	Boy 3	Boy 4	Boy 5
Girl 1	D	F	D	H	A
Girl 2	G	B	G	B	B
Girl 3	D	A	A	E	A

The letter D in the upper left–most box represents a problem solved by both boy 1 and girl 1. Problem D was also solved by girl 1 and boy 3 in common and by girl 3 and boy 1 in common. Because problem three states that each girl-boy pair solved at least one problem in common, Reid knew that all 441 boxes would contain a letter.

The other key piece of information was that each contestant solved at most six problems. Therefore, only six different letters could appear in any given row or column of 21 boxes. But think about the problem of distributing six letters among 21 boxes. If five of those six letters each appeared just twice in a row of 21 boxes, thus filling 10 boxes, the other 11 boxes would have to contain the sixth letter. In general, six distinct letters can be placed in the 21 boxes in a row only if at least 11 of those boxes contain letters that appear three or more times in that row.

Now Reid imagined going through each row and coloring red all the boxes with letters that appear at least three times in that row. For instance, the part of the table shown above has the B's colored in for girl 2, because the letter B appears at least three times. Applying the same rule to girl 3's problems would mean coloring in the A's in her row. Reid knew that at least 11 boxes in each row must be colored red, because of the constraints on distributing six letters among 21 boxes. And because the table has 21 rows altogether, at least 11 × 21 or 231 boxes must be colored red.

Reid then applied the same logic to the columns, except that he colored the boxes blue. In other words, in each column all the boxes with letters that appear at least three times in that column would be blue. By the same reasoning as for the girls' problems, at least 231 of the boxes had to end up blue.

But the table contains only 441 boxes. If at least 231 are colored red, and at least 231 are colored blue, then some of the boxes must be colored both red and blue. The letter in each of these doubly colored boxes represents a problem solved by at least three girls and at least three boys. Proving that such a problem exists solves problem three.

To a mathematician, watching the construction of a proof like this is best compared to watching someone compose a piece of music. The problem solver has to sense exactly which parts of a problem warrant attention. He or she has to grasp how the pieces of a proof cohere. The expenditure of effort needs to be balanced with the task, so that the subtle interplay of internal thoughts and the external challenge suggests what to do next.

Maybe that's also the best way to think about talent — as a musical composition. In both its development and its display, talent then would consist of parts played by the instruments of an orchestra. A part could be removed, but doing so would make the music incomplete, unbalanced. At times a single musical theme or instrument may rise to the surface, but soon the solo voice subsides, lost in the swirling counterpoint of sound.

6 · interlude:
An Afternoon to Rest

At exactly 1:30 P.M. the blare of the air horn again filled the cavernous interior of the Patriot Center at George Mason University. The 473 competitors put down their pencils. Some smiled ruefully at their papers and shook their heads, as if promising themselves never to be fooled by that trick again. A few others pushed angrily away from their desks. But most simply gathered the papers on their desk and placed them in the envelope that had held the questions. Slowly and stiffly they stood up.

Outside the arena the Olympians milled through the crowd, looking for their teammates. A few embraced, crying tears of anger or relief. Others shook hands, as they had before entering the arena. The assistant coaches and the guides had received the problems partway through the exam, but few had yet come up with solutions. Still, the team members clustered around the assistants and peppered them with questions. As competitors heard about steps they had overlooked, or approaches that seemed promising, or pivotal points in possible solutions, they groaned with frustration and asked how they could have been so stupid.

The American team was pleased but not overjoyed with its performance. Reid was confident that he had gotten perfect scores on all three problems. Ian knew he had solved problems one and two, but he had run into a glitch with problem three. The logic of his solution contained a small gap — which ultimately would cost him a point and keep him from getting a perfect score for the first day. Tiankai had aced problem one and had gotten very close to solving problems two and three, but in both cases his proofs were slightly flawed. He would have to hope for partial credit.

As each team made its way toward the cafeteria, they asked other teams, "How did you guys do?" And in this way the jockeying for position commenced. After the first day, teams have a distinct advantage in conveying the impression that they have done well. Until the judging is concluded, several days after the exam, no one knows the results. And it can be extremely discouraging for a strong team to hear from another team that the problems were easy, even if they weren't.

The Koreans were especially effective at this game. At lunch they casually mentioned that every member of their team had done well. This seemed unlikely, given the difficulty of the problems, but the news swept through the cafeteria in minutes — *the Koreans say the problems were easy.*

The Americans were actually more worried about the Chinese than the Koreans. The Chinese were sitting together at lunch, talking among themselves and seeming quietly confident. None of them knew much English, so information had to pass through third parties. But the buzz in the cafeteria was that all of the Chinese thought the morning had been an overwhelming success.

This was troubling news to the U.S. team members, who knew that they had lost a few points that morning. The U.S. team traditionally has been very strong but not dominant in the Olympiad. In the twenty-six events in which the United States had competed since 1974, the team had finished in the top five twenty-three times. But the United States had finished in first place just four times — in Yugoslavia in 1977; in 1981, the other time the Olympiad had been held in the United States; in Poland

and in Hong Kong in 1994, when everyone on the team had received a perfect score. Still, this was the strongest team the United States had fielded in years. Every member felt that the team had at least a chance to beat the Chinese, the Russians, the Romanians, and the other strong teams. And to finish first in the world on their home field — that would be a glorious triumph.

After lunch the Olympians wandered back to the dorms. The impromptu soccer game in the courtyard promptly resumed, though none of the American kids were playing — they didn't have much interest in soccer. The soccer players, shouting at each other in a dozen different languages, kicked the ball with a special vengeance.

Later in the afternoon Melanie Wood and the U.S. team members gathered in a common room overlooking the soccer game. They talked about the problems for a while and then fell silent. "What should we do?" someone asked.

"Let's play a game," Melanie suggested.

"Which game?"

"How about Twitch?" she replied.

Twitch must have been invented by a sadistic math Olympian. A deck of specially marked cards is dealt face-down to the players, who sit in a circle. When a player flips a card into the middle of the circle, the symbols on its face indicate who must play the next card. The card might read *R*, in which case the person one place to the right must play next, or *2L*, sending play to the person two places to the left. But some cards reverse the direction of play or make the game go in reverse. Twitch requires lightning-fast reflexes, because if you don't flip your card onto the stack before another player realizes you should play next, that person can force you to take all the cards in the stack. The first person to get rid of his or her cards wins the round.

The Olympians sat cross-legged on the floor, and the game began. The room was extremely warm, so someone opened a window. The shouts of the soccer players floated up from the courtyard. Some members of the Colombian team, sleeping on sofas in the common room, were so exhausted that the noise didn't wake them. A writer who had been following the U.S. team was one of the players. As he tried to keep up with the endlessly flashing cards, his head felt as if it were about to explode, so difficult was the struggle not to lose. He began to sweat, partly because it was so hot in the room, partly because he was falling farther and farther behind.

The kids on the team, on the other hand, seemed to get calmer as the game got more intense. They played their cards quickly, but their movements were concise, economical, controlled. While the cards were being dealt, they sat with a quiet concentration, waiting for the game to begin.

Reid Barton won almost every round.

7. creativity

All six members of the U.S. Olympiad team were superb pianists, and at the training camp one or another of them could often be found playing the piano in the common room. But all tended to stick to the classical music they had learned — except one. When David Shin sat down at the piano, he did not launch into a Beethoven sonata or a Chopin polonaise. Instead, he sounded out a melody — something overheard earlier that day, maybe, or a musical phrase running through his mind. Bending low over the keys, his eyes almost closed, he would add to the melody — a bass line, some harmonies. He would invert the notes and play them faster or slower, alter them slightly, or try a new accompaniment. Over and over he would take the melody apart and put it back together, until the other Olympians became exasperated with his noodling. David ignored them. "Just sitting at the piano is relaxing for me," he says. "If I'm doing math and need a break, I'll go play the piano for half an hour."

Rearranging things was the key to solving the formidable-looking problem that opened the second day of the Olympiad:

Let n be an odd integer greater than 1 and let $c_1, c_2, \ldots, c_n$ be integers. For each permutation $a = (a_1, a_2, \ldots, a_n)$ of $\{1, 2, \ldots, n\}$, define

$$S(a) = \sum_{i=1}^{n} c_i a_i$$

Prove that there exist distinct permutations b and c such that $n!$ divides $S(b) - S(c)$.

126

When people think about tough math problems, this is what they have in mind. Anyone without a solid grounding in mathematics will have a hard time figuring out even what the problem is asking. It features a complicated sum — denoted by the capital Greek letter sigma — that involves adding n pairs of numbers that have been multiplied together. It mentions permutations — different ways of ordering distinct objects. (In this problem a, b, and c represent not single numbers but permutations of the whole numbers from 1 to n.) And at the end the problem mentions the number $n!$, which is not a startled n but mathematical shorthand for the number n factorial, while is defined as $n \times (n - 1) \times (n - 2) \times \ldots \times 3 \times 2 \times 1$. Even among people who understand the mathematical intent of this problem, few would have any idea how to go about solving it. Yet David solved it with a single page of elegantly concatenated equations. As with all of the problems on the Forty-second Olympiad, problem four could be solved in several ways. The approach that David took was a model of mathematical rigor, concision, and creativity.

The kids on any U.S. Olympiad team differ greatly from one another. They are more or less outgoing, more or less athletic, more or less intuitive, even more or less "talented." But they all have one thing in common: an amazing mathematical creativity. The problems on any Olympiad cannot be solved by plugging numbers into memorized formulas. They require sidelong attacks, inspired guesses, flights of mathematical fancy. In that respect, the members of an Olympiad team are not only mathematicians: they are artists working in a medium of form and numbers.

▽

A scientific explanation of human creativity would seem to be a contradiction in terms. Science is the study of regularities in nature. Scientists observe events and try to conceive of underlying mechanisms that explain those events. They then test their hypotheses against their observations in an effort to develop well-grounded explanations for some aspect of our existence.

Creativity would seem to be the opposite of regularity. It seeks new perspectives on familiar surroundings, novel ways of interacting with the world, unexplored intellectual and emotional terrain. Even the means of achieving creativity can be approached creatively, as when artists, musicians, and even mathematicians began to use computers to generate new ideas and insights.

A traditional way of examining creativity has been to study the people who exhibit it. In the 1950s and 1960s psychologists at the University of California's Institute for Personality Assessment and Research in Berkeley compared highly creative and less creative individuals in a variety of fields, including mathematics. Highly creative people, the researchers found, tended to exhibit a distinctive constellation of traits. They were often iconoclasts who did not especially care what other people thought of them (though, paradoxically, many creators yearn for the approval of those whom they consider their peers). They tended to have strong egos and were often considered arrogant. They usually worked very hard in their chosen fields yet also had wide-ranging interests. Their internally imposed standards were very high, and they set challenges for themselves that forced them to struggle to achieve their goals.

Interestingly, researchers have found that creativity is not tightly associated with IQ scores. Some people who do well on IQ tests are not very creative, and some people who are highly creative do not score particularly well on the tests. It may be the case that a certain threshold of mental acuity is required for high levels of creativity — in his book *Creating Minds,* psychologist Howard Gardner sets the threshold at an IQ of about 120. But some creative people, such as artists with savant syndrome, have IQ levels well below average.

Correlations between creativity and character are intriguing, but they don't tell us much about the nature of creativity. Some creative people exhibit few of the traits psychologists associate with creativity. And, as with all correlations, the direction of causality is uncertain. Do these traits cause people to be creative, or do creative people tend to develop these traits?

Dean Keith Simonton, a psychology professor at the University of California at Davis, has been thinking about creativity for much of his career. In the early 1970s, his Ph.D. work at Harvard involved charting the ebb and flow of creativity throughout history. He compiled vast lists of eminent creators and their accomplishments and found that the overall creativity of a culture hinges on factors that may seem only distantly related to the lives of isolated poets, composers, and mathematicians. For example, political openness and pluralism seem to increase the amount of creativity in a society by exposing potential creators to a variety of cultural perspectives. A good example, says Simonton, is classical Greece, which entered a period of tremendous creativity after winning the Persian wars and beginning a series of intensive trading interactions with its Mediterranean neighbors. Similarly, when a formerly closed society opens to outside influences, a golden age of creativity can ensue. In a study of Japanese history, Simonton found that creativity soared when Japanese students studied abroad, when the nation interacted with foreign powers, or when it permitted individuals from other countries to immigrate. When Japan's rulers limited foreign influence and travel abroad, creative momentum stalled.

In recent years Simonton has been investigating the broad range of factors that come together to produce notable individual achievements. He has studied the productivity of classical composers, the success of U.S. presidents, independent discoveries in science, and eminent individuals in minority cultures. He has decided, after this wide-ranging examination, that the best

way to think about creativity is to compare it to a different phenomenon: biological evolution. In nature, evolution occurs
through a three-step process. The first step is the generation of
novelty. Organisms are born with traits different from those of
their parents and siblings because every organism has a unique
combination of genetic variants (even clones have a few genetic
differences) and undergoes unique experiences as it develops.
The second step is selection. Some organisms thrive because they
have traits that are useful in their environment, while less advan-
taged organisms falter. The third step is replication. Organisms
that thrive tend to have more offspring than do their less successful siblings, so they have more opportunities to pass on their advantageous traits. Over time this process gradually produces organisms that are increasingly well suited to the physical,
biological, and cultural environments in which they live.

Simonton believes that creative acts — the production of
something that has never been produced before — occur
through a similar three-step process. The first step is the generation of new ideas. In Darwinian evolution, new biological traits
arise through essentially random processes as genes mutate, re-
sort themselves, and interact with the environment in each new
generation. The origins of new ideas can be equally random, according to Simonton. The mind takes apart old ideas and observations and juxtaposes their elements in new ways. The process
is "unpredictable and chaotic," he says. "You have to try lots of
things and go up lots of blind alleys. You generate a whole bunch
of ideas that are loosely connected, because you don't know in
advance whether they're going to work or not."

In his book *Origins of Genius: Darwinian Perspectives on
Creativity,* Simonton quotes the psychologist William James's description of the creative ferment:

Instead of thoughts of concrete things patiently following
one another in a beaten track of habitual suggestion, we

have the most abrupt cross-cuts and transitions from one
idea to another, the most rarefied abstractions and discrimi-
nations, the most unheard of combination of elements, the
subtlest associations of analogy. . . . [W]e seem suddenly introduced into a seething cauldron of ideas, where everything
is fizzling and bobbling about in a state of bewildering activity, where partnerships can be joined or loosened in an
instant, treadmill routine is unknown, and the unexpected
seems only law.

The elements of a new idea may be words, images, melodies, equations, emotions, snippets from dreams — anything the
mind can imagine or feel. When Tiankai was solving problem
one, for example, he had the image of a 30-degree angle in his
mind. When he saw how the angle could be superimposed on the
triangles he had drawn, he had all but solved the problem. In
cracking problem two, Ian thought of Jensen's inequality and
suddenly discovered a particularly elegant solution.

Sometimes this juxtaposition of ideas occurs consciously.
Some visual artists try to overlay disparate images to loosen their
preconceived ideas about how things should look. Or a creator
tries to hold two contradictory ideas in his or her mind simultaneously to generate a cognitive tension that produces new insights. Mathematicians sometimes use similar tricks to spur creativity. In solving problem two, for example, Ian might have
consciously gone through a list of possibly relevant memorized
equations.

But more often the process seems to occur subconsciously,
which is how Ian said he arrived at his solution. You're returning
from the bathroom or walking along a path or boarding a bus,
and suddenly the solution is there, elaborate and complete, as if
bequeathed to you by a kindhearted muse.

A classic experiment that has been used for decades to study
creativity was developed by the psychologist Norman Maier at

the University of Michigan in the 1930s. A subject is led into a room that has two strings hanging from a high ceiling. He is told that the task is to tie the two strings together. The strings are long enough to be tied, but if the subject grabs the end of one and pulls it toward the other, the end of the other string always remains just out of reach. How can the problem be solved?

In the original experiment various objects were scattered around the room, and the subject was told that any of them could be used in the solution. But only one of the objects, a pair of pliers, is actually useful. By tying the pliers to the end of one string, the subject could start that string swinging like a pendulum. He or she then could hold the other string and, when the pliers swung within reach, grab the pliers and the connected string, remove the pliers, and tie the strings together.

Many people have great difficulty solving this puzzle. But the experimenters found that certain clues could make it easier. For example, if the pliers were replaced by a plumb bob, more people figured it out, evidently because the bob suggested something that should be tied to the end of a string. Women did better at solving the problem if the pliers were replaced by a scissors.

But the most interesting variant involved the subconscious mind. If the investigator doing the experiment walked into the room and "accidentally" brushed up against one of the strings so that it started swinging, the subjects were much more likely to solve the problem. When asked how they arrived at the solution, however, they usually failed to recognize the swinging string as a clue. They said the solution just popped into their minds.

Many scientists, mathematicians, composers, writers, and other creative artists have reported moments of epiphany when the solution to a problem suddenly materialized in their thoughts. Charles Darwin experienced such a moment, suddenly realizing how the selection of organisms with advantageous traits would lead to the origin of new species. "I can remember the very spot in the road, whilst in my carriage, when to my joy the solution occurred to me," he wrote in his autobiography. Such moments are so startling and wonderful that many creative people have developed techniques to kick-start the process. David Shin plays the piano. Many creators go for long walks. A moment of rest sometimes can free the mind. On the first day of the Olympiad, Ian read through the problems, put his head on his desk, and took a short catnap. A good night's sleep can be even more productive. "I remember waking up knowing the solution to a problem," David says. "I was working on it the night before, a number-theory problem. I couldn't get it, so I went to sleep. When I woke up the next morning I thought about it for a moment and the answer was right there."

Many creative people have learned how to relax and let their minds wander so that their subconscious is freed to generate ideas. When you focus your entire attention on a problem, your mind may dwell exclusively on what it already knows, leaving no room for new ideas. "One of the main predictors of creativity is being open to what is happening around you," says Simonton. "You're an open system rather than a closed system. Maybe you hear something that isn't immediately relevant, but it sets your mind working in a direction in which you weren't going to go. That kind of flexibility and openness is crucial for successful problem solving."

If creativity does require the novel juxtaposition of ideas, that might help explain some of the traits exhibited by highly creative people. Their defiance of convention and authority might free their minds to consider new combinations of ideas. Though they are able to focus intently on problems for long periods, their wide range of interests may suggest unusual approaches to a problem. Studies have shown that creative people often are familiar with more than one culture or language and that they tend to value complexity, diversity, and different viewpoints. "They have a tolerance of ambiguity," says Simonton, "which allows them to tolerate the fact that they're not going to find a solution right away. What keeps a lot of people from being creative is that they're too impatient. They want a solution fast. And often they want a solution that meets certain *a priori* specifications, even though those *a priori* specifications may rule out the only solutions that are possible. As a consequence, they can't generate the various combinations of ideas that they need to find those particular combinations that are most likely to lead to a solution."

All of the math Olympians were experts in generating ideas that might prove useful. Even when they were not doing math, they often seemed to be exercising their creative faculties. At the training camp, for example, they rarely played regular chess, which they considered too boring, but they incessantly played chess variants. In the game known as suicide chess, the object is to be the first player to lose all your pieces. In atomic chess, whenever a piece is captured, so are all the pieces on adjacent squares. The board in toroidal chess is connected from side to side and from front to back, so a piece can move off one side of the board and reappear on the other side. In proxy chess each piece has the moves of the piece to its immediate left. And the Olympians were especially fond of bughouse, a devilishly complex game played against the clock with two chessboards, in which captured pieces immediately reenter the game on the other board.

▽

The idea that creativity begins with the novel juxtaposition of ideas immediately runs into one major objection. As soon as you consider more than a few discrete items, the number of ways of combining those items becomes very large. The number of distinct permutations of n different things is n factorial or $n!$ — the same number that comes up in problem four. This number is manageable when n is small. For $n = 3$, $n!$ is $3 \times 2 \times 1 = 6$. For $n = 4$, $n!$ is 24. But say that you're combining ten things in different ways. The number of distinct permutations of those ten things is 10 × 9 × 8 × 7 × 6 × 5 × 4 × 3 × 2 × 1 or 3,628,000. That's why a sentence exactly the same as this one has almost certainly never before been written in all of human history. The twenty-two words in that sentence can be recombined in more than one sextillion ways (1,124,000,727,777,607,680,000 ways, to be exact). Only a tiny fraction of those combinations will make any sense, but even this small fraction must amount to many hundreds or thousands of different sentences. The explosion of possible combinations is also the reason songwriters will never run out of melodies — and they have the extra advantages of being able to vary the duration and volume of each note and the instrument on which a note is played.

With so many ways of combining things, creators cannot consider every possibility. They must go through a selection process by which they first consider the most promising options. This is the second of Simonton's three steps in creativity. "To some extent, you preselect the range of elements you're going to consider," Simonton says. "But it's a broad selection process, because you're saying that out of the one hundred possible ways to solve a problem, these dozen are the best. Then, when you find yourself in a situation where you can't solve the problem with those dozen, you start broadening your scope."

The game of chess offers a good example. When expert players face a given situation in a game, they don't think of every possible move they could make. They consider a range of moves that are most likely to be effective in those circumstances. That's one reason why grandmasters sometimes have so much difficulty playing against computers, explains Simonton. Computers can consider a broader range of possible moves than can humans, including moves that human players would immediately discard as useless. But every once in a while one of those seemingly useless moves is actually very effective. "When you play a human being, you know the range of strategies they're going to use," says

Simonton. "But computers will sometimes come out of the blue with a strategy that you've never seen before. And then you're put in a situation where you have to consider all possible moves, which is something you don't normally need to do as a chess player."

The process of selecting ideas from a range of possibilities is inevitably idiosyncratic. It depends on many aspects of a person's personality and varies widely among individuals. People use their own knowledge, previous experiences, foresight, intuition, aesthetic judgments, and blind luck to make decisions. Sometimes they use all at once, blending the rational and irrational in proportions difficult to discern.

The idiosyncratic nature of creativity is apparent in an Olympiad. Each competitor has a slightly different way of analyzing a problem and looking for a solution. A visually creative thinker such as Tiankai might treat problem four spatially, converting the numbers from 1 to n into a ring and matching them with other numbers. Or someone like Reid might place the numbers in a grid, as if they were being analyzed by a computer. This choice is in part an aesthetic one, a matter of personal preference.

One interesting way to think about how ideas are selected is to look at individuals in whom the process is to some extent broken. Throughout history, commentators have speculated about the link between creativity and mental illness. "No great genius has ever existed without some touch of madness," wrote the Roman philosopher Seneca in *On the Tranquillity of the Mind.* "Genius is to madness near allied / And thin partitions do their bounds divide," stated the poet John Dryden. "I have long had a suspicion," wrote the British psychiatrist Henry Maudsley in 1871, "that mankind is indebted for much of its individuality and for certain forms of genius to individuals [with] some predisposition to insanity. They have often taken up the bypaths of thought, which have been overlooked by more stable intellects."

From Simonton's perspective a link between madness and creativity makes sense. A touch of madness may enable creators to combine ideas in unexpected ways. Or the range of ideas may be less constrained in people whose brains travel little-used paths. Consider the writing of Philip K. Dick, who was born in Chicago in 1928, moved to California as a child, and often seemed to veer close to madness in his personal life. In his 1981 novel *Valis,* Dick wrote a long appendix that contains writing like this passage: "45. In seeing Christ in a vision, I correctly said to him, 'We need medical attention.' In the vision there was an insane creator who destroyed what he created, without purpose; which is to say irrationally. This is the deranged streak in the Mind; Christ is our only hope, since we cannot now call on Asklepios."

And so on through references to Zeus, Apollo, Elijah, Pascal, the Empire, the hologramatic universe, homoplasmates, the divine syzygy, to the final sentence of the book: "But underneath all the names there is only one Immortal Man, and we are that man." Dick died in obscurity and poverty in 1982, though since then his work has undergone a revival (the movies *Blade Runner, Total Recall,* and *Minority Report* were all based on short stories by him). He said that he wrote *Valis* and two related novels after a mystical experience in March 1974 that he described as "an invasion of my mind by a transcendentally rational mind." Yet few people would consider Dick's work rational. If an editor received one of his manuscripts without knowing the author, it almost certainly would be relegated to the dusty pile of submissions from lunatics.

In mathematics a well-known example of a possible link between madness and creativity can be found in the work of the mathematician John Nash at Princeton. Nash tried to approach familiar problems from unconventional directions. He tended not to read deeply in the mathematical literature because he did not want to be distracted by other people's ideas. According to his biographer Sylvia Nasar, his "flashes of intuition were non-rational" at the same time that he was a "compulsively rational" person. She wrote, "A predisposition to schizophrenia was probably integral to Nash's exotic style of thought as a mathematician."

These examples and others like them are suggestive, but they can't be the whole story. For one thing, most mental illness is incapacitating rather than inspiring. When Nash descended into full-blown schizophrenia shortly after his thirtieth birthday, he was unable to continue doing creative mathematical work. The selection process in his mind seemed to lose its grip entirely, so that he thought he saw secret messages in newspaper headlines or in people's neckties. At that point his mind really did seem to be considering all 3,628,000 possible combinations of ten items.

Furthermore, most eminent creators — including mathematicians — exhibit no signs of mental illness. On the contrary, many are very well adjusted, which accords with another view in psychology: that creativity is the natural product of a balanced, self-fulfilling personality. Certainly none of the Olympians on the team demonstrated any sign of mental problems, and it is difficult to think how they could. The tremendous competitive pressures of the qualifying process would almost certainly weed out any but the most resilient personalities.

In his groundedness, David was typical of all the team members. Growing up in West Orange, New Jersey, he was more interested in music than math. "I play the piano, drum set, French horn, baritone horn," he says. "I played in every musical group the school offered — jazz band, marching band, pit orchestra, brass ensemble." Only after finishing with a high score in Mathcounts during middle school did he decide to devote himself to becoming a better problem solver. "Most of the math I've learned I've learned on my own," he says. "I've never had a coach. My father could help me up through Mathcounts, but after that I was on my own. The thing I've most appreciated about my parents is that they've never pushed me. I have a friend who used to be real good when I was in seventh grade. When I got first in the school in Mathcounts, he got second. But his parents pushed him so much that pretty soon he didn't care. My parents let me discover my love of math on my own. I'm really thankful for that."

▽

Generating and selecting novel ideas are essential steps for creativity. But they mean nothing without what Simonton sees as the third step in the creative process. A new idea has to encounter a receptive environment so that it spreads from mind to mind. Otherwise it dies with the individual. (This aspect of creativity applies more to mathematical research than to competitive mathematics, in which the solution to a problem is already known to exist. In competitive math, replication often builds on success.)

The ways in which new ideas are replicated can be as idiosyncratic as their generation and selection. Sometimes people immediately recognize the value of an innovation. To use a technological analogy, once color televisions were developed everyone saw that they were an obvious improvement over black-and-white sets. As they became more affordable, they replaced black-and-white televisions almost everywhere. In evolutionary biology, this kind of rapid replacement of one variant with another is called a selective sweep.

Sometimes an idea takes longer to gestate. To take another technological example, portable phones first became available in the 1960s and 1970s. But the first ones were bulky contraptions more like radio receivers or walkie-talkies. Gradually they became smaller and cheaper. At the same time, people became more comfortable with the idea of talking on a phone while driving a car or eating dinner in a restaurant (maybe too comfortable), so their use spread.

A third possibility is that an idea is ahead of its time. The facsimile machine, for example, was invented more than a century and a half ago, in 1842. But its use was severely limited because it was necessary to have compatible machines on the sending and the receiving ends. Only when a widespread network of fax machines emerged in the 1980s did faxes go from being a rarity to commonplace.

The replication of ideas depends on so many factors that predicting whether a new idea will succeed seems impossible. But one generalization seems secure. Only if an idea is born into a cultural environment that needs or desires it will it succeed, although the desire or need may not become apparent until the idea exists. The creator who is not addressing a need within society will remain unknown and unheralded. Creativity therefore depends on much more than the individual creator. It takes place as an interaction between an individual's capabilities and the social environment. As the psychologist Mihaly Csikszentmihalyi has put it, the critical question is not "What is creativity?" but "Where is creativity?"

The social dimension helps explain why bursts of creative activity occur at particular points in history. Circumstances arise that encourage innovators and produce receptive outlets for their ideas. Examples include the development of monumental architecture in ancient Egypt, the classical period in music from Bach to Beethoven, the writing of the U.S. Constitution after the American Revolution, impressionist painting in France at the end of the nineteenth century, the creation of quantum mechanics in the first half of the twentieth century, and the worldwide popularity of American movies at the close of the twentieth century. In each of these cases, innovations arose through the actions of individual creators. But an equally strong influence was the receptiveness to those creations (even if somewhat delayed) of the broader society.

This view also can help explain why bursts of creativity sometimes occur within particular groups, including particular ethnic groups. Creators usually learn their craft from people whom they admire and emulate. These models and mentors are often members of their own social group, and thus particular groups can become known for certain kinds of achievements. Think of blues music among African Americans, chess among eastern Europeans, winemaking in France, or engineering in Japan.

Finally, the social dimensions of creativity offer a somewhat different perspective on the products of "genius." Great accomplishments do not exist in a social vacuum. They are created through an interaction between a creator and the individuals who perceive and acknowledge that work as extraordinary. In this sense the members of a society actively participate in the creation of works of genius. No matter how isolated or detached from reality a creator may be, the link between that person's work and the broader society must remain strong for a new creation to survive and prosper.

▽

Maybe these observations can help answer a question asked by almost everyone who spends much time at U.S. math competitions. Why are so many of the competitors of Asian ancestry? The U.S. team at the Forty-second Olympiad was fairly typical of recent teams. Tiankai's family was from China, Ian's family was from Vietnam, and David's family was from Korea. In recent years about half the kids at the training camp and on the team have had Asian backgrounds, even though Asian Americans, including people with ancestors from Pacific islands, make up only about 3 percent of the U.S. population.

Questions about the academic achievements of Asian Americans are not limited to math competitions. The group has a reputation as a "model minority" that excels academically. Asian Americans are overrepresented in gifted and talented classes from elementary school through high school. Compared with all other ethnic groups, including European Americans, Asian Americans have higher rates of graduation from high school, college matriculation, and graduation from college.

One possible explanation is that people with Asian ancestors are biologically smarter. That is, maybe their brains are put together differently so that math, science, and other academic subjects are easier for them. Some fringe biologists have even concocted evolutionary just-so stories that seek to explain such capabilities. They say that the ancestors of today's Asians had to be smarter than other people to survive on the freezing cold plains of Asia during the Ice Age. Or that success on civil service exams (in China, at least) selected for intelligent people and enabled them to have more children than others.

This idea of Asian mental superiority has many problems. First, exactly who qualifies as an Asian? Do Asian Indians, who often — but not always — do well at Olympiads? Most of their ancestors lived in warm climates and never took civil service exams. How about central and northern Asians like the Mongols and Siberians? They lived in even colder climates than did the ancestors of the Chinese. And what about Native Americans, who are descended from Asians who migrated to the Americas during the height of the Ice Age?

Also, the claim that everyone with Asian ancestors does well in mathematics and science is obviously overblown. Among Asian American students, only a small percentage excels in mathematics and science, just as with students belonging to other ethnic groups. And the children of some Asian ethnic groups, especially groups disadvantaged in Asia, have many of the same academic difficulties as do other disadvantaged groups. Some commentators have concluded, after careful reviews of census and other data, that the model-minority stereotype is as misleading and as counterproductive as other ethnic stereotypes.

But the most convincing evidence comes from direct comparisons of how different groups of children actually think. In the 1980s psychologist Harold Stevenson and his colleagues at the University of Michigan and in Japan and Taiwan studied groups of several hundred first-graders and fifth-graders in each of three cities: Sendai, Japan; Taipei, Taiwan; and Minneapolis. The researchers constructed a battery of tests carefully designed to detect differences in thinking abilities, not just differences in teaching methods or materials. The tests asked students to remember a sequence of tones, match spatial patterns, recall lists of words or numbers, answer questions about a brief story, solve mathematics problems, and so on.

Certain groups of children did somewhat better on particular tasks. The Chinese children were better able to remember lists of numbers than the Japanese and American children, the Japanese kids did best at remembering tones, and the American kids did best at matching shapes. Taken together, however, the differences among the groups were small — a conclusion confirmed by other tests of cognitive abilities conducted since then. Stevenson and his colleagues wrote: "The results suggest that the high achievement of Chinese and Japanese children cannot be attributed to higher intellectual abilities, but must be related to their experiences at home and at school."

If experiences make the difference, then what are these experiences? After all, if they make a difference for Asian American kids, they should work for everyone else. Stevenson and his colleagues looked at this question in a series of follow-up studies and came to a strong conclusion. The critical factor in student achievement, they said, is parental attitudes. American parents typically are content with their children's performance in school, even when that performance is far below international standards. Among the parents of U.S. eleventh-graders interviewed in 1990, almost half said they were "very satisfied" with their child's academic achievement. The percentages of Chinese and Japanese parents reporting a comparable level of satisfaction were in the single digits.

In general, Chinese and Japanese parents stress the importance of working hard to succeed. They expect their children to do well in school and get good grades. Children learn from an early age to respect and defer to authority figures such as parents and teachers. They are told that if they do poorly in school, their performance will reflect badly not just on them individually but on their whole ethnic group.

Surprisingly, however, Stevenson has found that the stereotype of overstressed Asian students driven relentlessly by their parents to succeed is not accurate. On the contrary, his surveys have found that American students are more likely to be stressed, depressed, or unable to sleep because of academic pressures than are Japanese or Chinese students. The majority of American students and families tend to believe that mathematical ability is innate; either you are born good at math or you aren't. American schools tend to track students into different curricula, with the more advanced kids in one class and less advanced kids in another, whereas in Asian schools everyone takes the same math classes. America is supposed to be the land of opportunity, where people are free to make their own futures. Yet the idea that individuals are stuck with an inborn set of talents seems to weigh heavily on many U.S. schoolchildren.

Many Asian Americans share an ethnic culture that has other distinct features. The children are much more involved in musical activities than are the children of other ethnic groups. Also, Asian American kids are less likely to be on a sports team, to participate in extracurricular activities, or to have a job outside school — they are expected to devote their free time to schoolwork.

These findings are suggestive, but another line of analysis points toward even deeper forces. Ethnic groups in America define themselves partly by the stories they tell about their strengths and weaknesses. Some of these stories center on sports, some on business, some on entertainment. Many of the stories Asian Americans share, both within their families and among friends, involve science and mathematics. Most Asian American children don't see themselves growing up to be NBA players, captains of industry, or politicians; they tend to believe that U.S. society functions in such a way as to cut them off from such options. But many believe that if they do well in mathematics and science, they can succeed. They can become scientists, engineers, computer programmers, physicians. Joining a math club and participating in math competitions is a way of reinforcing these internal narratives. Relatively few African American or Hispanic kids would say, if asked, that they want to be one of the best high school mathematicians in the United States, while thousands of Asian American kids would respond that way.

Then again, maybe there's no need for complex psychological explanations for the overrepresentation of Asian Americans in math competitions. Think about the three Asian Americans on the team representing the United States at the Forty-second Olympiad. All were born outside the United States. Tiankai came from China; Ian was born in Australia, though his parents had emigrated from Vietnam; and David had emigrated from Korea.

People who come to the United States as children cannot help but be deeply affected by the experience. They grow up hearing about or experiencing the stories of struggle and perseverance required to succeed in America. Many Asian American kids are from relatively privileged families; otherwise they would not have been able to leave their countries. But all new immigrants to the United States must work hard to succeed, and they expect their children to work hard, too.

With each new generation born in the United States, the immigrant experience fades. By the third generation most Asian American kids are more American than Asian. First-generation immigrants from Asia tend to receive grades at school that are higher than the average, but over the generations the grades regress to the mean. On many measures of health, attitude, and well-being, recent immigrants score far higher than families that have been in the United States for longer periods.

The ethnic makeup of U.S. Olympiad teams clearly shows this effect. Most Japanese families in the United States, for example, have been in the country since before World War II. As third- or fourth-generation Americans, most of the young people no longer speak Japanese. They tend to be good students, but they do not necessarily excel in mathematics, and they do not gauge their self-esteem in those terms. Accordingly, they are not particularly numerous at math competitions, and no U.S. Olympiad team has included a member with a Japanese background.

Most of the kids from China, Vietnam, Korea, and other Asian countries, on the other hand, are from families that have emigrated more recently. They speak more than one language and have experience with multiple cultures, which, as Simonton demonstrated, can be a source of creativity. From an early age they absorb the lesson that they must work hard to do well in the United States and that if they master mathematics and science they are more likely to succeed. Given the precarious position of immigrant families in U.S. society, the intensity of their drive to succeed is hardly surprising.

▽

Simonton believes that many interacting factors must come together to result in any creative act, whether a symphony, a poem, or a solution to problem four on the International Mathematical Olympiad.

The problem asked the Olympians to prove a particular property for the complicated sum

$$S(a) = \sum_{i=1}^{n} c_i a_i$$

The sigma means adding a sequence of terms. Thus, another way of writing $1 + 2 + 3$ is

$$\sum_{i=1}^{3} i$$

Problem four says that a set of integers is represented by the terms $c_1, c_2, c_3, \ldots, c_{n-1}, c_n$, while a permutation of the first n whole numbers is represented by $a_1, a_2, a_3, \ldots, a_{n-1}, a_n$. The first member of one set is multiplied by the first member of the other set (so $c_1 \times a_1$), then the second members of the two sets are multiplied ($c_2 \times a_2$), and so on through $c_n \times a_n$. Then all of these products are added together, giving $c_1 a_1 + c_2 a_2 + \ldots + c_n a_n$, and the result is $S(a)$. Thus, the a in $S(a)$ represents a permutation or unique ordering of the whole numbers from 1 to n. Since the first n whole numbers can be arranged in $n!$ different ways, the term a can take $n!$ different forms, as can the sum $S(a)$.

Many problem solvers would quickly identify the sum as the most complicated part of the problem and therefore would try to avoid it. David immediately realized that the only way to solve the problem was to attack its most difficult part. Essentially, he would have to make the problem more complicated before he could make it less complicated.

He decided to take the sum of all the possible values of $S(a)$ for the $n!$ different values of a. Mathematically, that produces a monster that looks like this:

$$\sum_a \sum_{i=1}^n c_i a_i$$

Having defined this sum, David proceeded by contradiction (as described in the appendix). He assumed that the statement he wanted to prove was false, then showed that such an assumption allowed the above "sum of all sums" to be calculated in two contradictory ways. The statement therefore had to be true, and the problem was solved.

The solution was masterful, but David had a hunch from the start that his approach would work. He knew from previous experience that a sum like this one has to be tamed for the problem to be solved. "When you're dealing with a permutation, it's hard to describe each individual one, because they're all different," he said. "But when you sum over all the permutations, you get rid of the variability." In other words, the key to solving the problem was not to be intimidated by its complexities. When David refused to back down, the variability represented by each permutation yielded to a higher degree of order.

David could not have solved this problem without a deep knowledge of mathematical traditions, which highlights one last dimension of creativity. Any new idea, despite its uniqueness, builds on a history of thought and experience that extends far into the past. Biological evolution works the same way. In nature the equivalent of a new idea is a newly born organism. It partakes of a biological history that extends into the distant past, yet if that new organism has the right characteristics, or the right experiences, or is just plain lucky, it can change the world. As Ralph Waldo Emerson wrote, "The creation of a thousand forests is in one acorn."

8 · breadth

A few days before the Olympiad the members of the U.S. team appeared on the television show *Good Morning America*. All six were crammed onto a tiny set in Washington along with two adults: coach Titu Andreescu and Tom Leighton, the chief scientist of Akamai Technologies, one of the major funders of the event. The producers had decided before the interview that the show's hosts would ask one team member a math question that he hadn't heard before. But there had been some confusion over who would answer the question. Most of the team members thought it would be Oaz Nir, who, with his casual good looks and easygoing nature, had gradually emerged as the spokesman for the team. But no one had told Oaz about the plan. He thought the question would be posed to Gabriel Carroll, one of the team's strongest and quickest mathematicians.

So when one of the interviewers said, "Here's a question that I understand Oaz is going to answer. How can you use a nineteen-degree angle to construct a one-degree angle?" the situation had all the makings of a disaster. Here on national television, with his family and his friends — the whole world — watching, as the camera zoomed in on him and the others waited expectantly, Oaz could have made a complete fool of himself. For the briefest instant a cloud passed across his face. Then he said, "Well, you could take the nineteen-degree angle and use it ten times to make a one-hundred-ninety-degree angle. Then you could remove one hundred eighty degrees from that angle using a straightedge, leaving you with an angle of ten degrees. You could double that angle to twenty degrees and then remove nineteen degrees, and you'd be left with a one-degree angle."

The hosts of the show had absolutely no idea what Oaz had just said. Turning to Tom Leighton, they asked, "Is that right?" Leighton replied, "Yes, it is." With obvious relief, the hosts returned to more comfortable questions.

Very few people could have responded to that question as adroitly as Oaz did. Most people have forgotten how to construct angles with a compass and straightedge, so they wouldn't know where to begin. And even many professional mathematicians would be flustered by the unusual setting, the unblinking television cameras, the pressure of the moment. Oaz's composure said something important about the math Olympians. They have other resources to draw on when their mathematical knowledge is not enough. These resources can vary: one Olympian might be a good improviser, while another has a steely determination. But each has distinct nonmathematical skills, and these skills have a big influence on what they do with their lives when they grow up.

▽

That Oaz would become the kind of person who could act as spokesman for the team could never have been predicted just a few years earlier. One of his seventh-grade teachers remembers him as "very humble and soft-spoken. There wasn't a boastful thing about him. Everyone always looked up to Oaz, because he was such a talented student. But as far as being a leader, I never saw that."

He was born in New Orleans in 1983, a few years after his parents immigrated to the United States from Israel. (His parents' background "wrecked my chances for a southern accent," he says.) Before he was old enough to begin school, his father's career as an engineer took the family to Jackson, the state capital

of Mississippi. Jackson, the metropolitan hub of an overwhelm-
ingly rural state, has risen from the forests and fields of central
Mississippi to become a place where people go to shop, eat, and
visit the science museum. But it retains at least some of the agrar-
ian tranquillity and isolation of its surroundings, despite all the
new Starbucks outlets and Target stores.

Oaz attended public elementary schools through the fifth
grade. But his parents, recognizing his talents as a student, realized that he needed more of a challenge. In the sixth grade he
entered St. Andrew's Episcopal School, one of the most academically rigorous private schools in the Old South. Founded in
1947, its motto is *Inveniemus Viam Aut Faciemus* — "We will
find a way, or we will make one."

The middle and upper schools of St. Andrew's are located in
a wooded area just north of Jackson, where the "pines in opening
vistas splashed with fading dogwood," in William Faulkner's description, are rapidly being bulldozed to make way for four-
thousand-square-foot homes. At the entrance to the school is a
small lake with a wooden cross on one shore. On the other shore
is a small but obviously sophisticated astronomical observatory.
At the recent opening of the school's new five-hundred-seat the-
ater, the Mississippi Symphony Orchestra played to commemo-
rate the event.

From the beginning, Oaz was an exceptional student. "He
was a good writer, interested in history," says Pam David, who
was his sixth-grade teacher that first year and is now head of the
middle school. "He was a good citizen at this school." But his
talents shone especially in math. "We had to think of things to
keep him busy," says Marcia Whatley, who taught Oaz's math
class in the sixth grade. "He taught me as much as I taught him."

By the seventh grade, Oaz's teachers were more or less letting him study what he wanted in mathematics, so long as he
kept up with the material being covered in class. But that year

St. Andrew's hired a teacher named Barbara Cirilli (her married name is now Tompkins) and asked her to organize a Mathcounts team. "That was a tremendous opportunity for me," says Tompkins, who now teaches at a huge public high school a few miles down the road from St. Andrew's. "I came into a school that had these tremendously talented kids, and the school encouraged me to do as much as I could with them."

In seventh grade Oaz finished first in Mississippi in Mathcounts, and Barbara Tompkins became the coach of the state team. In May the two of them flew to Washington for the national competition. There, for the first time in his life, Oaz met other accomplished young mathematicians. It was a deflating experience. "There are fifty-seven teams at the national Mathcounts competition, so there are two hundred twenty-eight people," Oaz recalls, "and I was exactly at the halfway point; I was one hundred fourteenth. So that's fairly mediocre."

He returned to St. Andrew's determined to improve. Tompkins went through her shelves and pulled out problem-solving books that she gave to Oaz. Among them was *The Art of Problem Solving,* the same book Melanie Wood had read with such enthusiasm after her success in Mathcounts a few years earlier. Slowly and carefully, sometimes with help from his father or Tompkins, but more often on his own, Oaz worked his way through the books, solving as many problems as he could. The next year, during eighth grade, he was eighteenth at the national Mathcounts competition. "That's not great, but it's a lot better," he says.

As with the other Olympians, one of Oaz's distinguishing characteristics as a problem solver was his perseverance. Tompkins lent him more books, and Oaz began reading the high school and college mathematics texts in the St. Andrew's library. "It was phenomenal to see how much he improved," says Tompkins. "He excelled at whatever he set his mind to do. I just provided him with the books and with the opportunity to compete. He did all the rest on his own." According to Oaz, "I didn't really get good at math until I started working really hard at it."

As a ninth-grader at St. Andrew's, he did extremely well on the American High School Mathematics Examination. He was invited to take the USAMO, in which he received an honorable mention, meaning that he was one of the top two dozen in the nation. As a result he was invited to attend the summer training camp in Lincoln. "I was very excited about going," he says. "Back then the people at the camp were like celebrities to me."

That summer was fateful for another reason. Because of his father's job, the family moved again — to the San Francisco Bay Area. His parents carefully scouted the schools in the towns where they could live. In the end they decided on Cupertino — a town of about fifty thousand people midway between Palo Alto and San Jose — where Oaz and his older brother could attend Monta Vista High School.

The school, which occupies a ramshackle collection of low-rise wood-frame buildings, nestles up against the soft burr of the coastal range. It faces east, toward the rising sun and the milky haze that often overlies San Francisco Bay. Cupertino is in the heart of Silicon Valley, and the road to the school passes buildings filled with software and biotechnology companies: Luminous Networks, Celerity Digital Broadband, Endotex, Lepton Networks, In-Time Software. Many hard-driving Silicon Valley executives live in Cupertino so their children can attend Monta Vista. It was a California Distinguished School in 1996 and a national Blue Ribbon school in 1998, and it is generally considered one of the top ten public schools in California. Newspapers in China and Japan routinely tout the school as a good place for immigrants to send their children.

Monta Vista's reputation and the intensity of many Cupertino parents have made the school an academic hothouse. "Our biggest problem here is students who are stressed out because they're not doing well academically," says one of the school's teachers. Still, Monta Vista is a public high school, so it includes all types of kids. "Remember high school?" writes David Brooks in the *Atlantic Monthly.* "There were nerds, jocks, punks, bikers, techies, druggies, God Squadders, drama geeks, poets, and Dungeon & Dragons weirdoes. All these cliques were part of the same school: they had different sensibilities; sometimes they knew very little about the people in the other cliques; but the jocks knew there would always be nerds, and the nerds knew there would always be jocks. That's just the way life is."

The range of cliques may be less broad at Monta Vista than elsewhere, but Oaz found that the school had an energy and diversity that he hadn't known at St. Andrew's — and he liked it. "Between the time when Oaz arrived here as a sophomore and when he graduated, he became a completely different person," says Bob van Hoy, a computer science teacher who befriended Oaz during that first semester. At first Oaz kept to himself a lot. He has always thought of himself as shy, despite his appealing smile, friendly demeanor, and obvious competence. But at Monta Vista he began to try new things, and he succeeded in them just as he had in math. Back in Jackson he had been on a swim team. Now he joined the school's water polo team, converting those endless hours of staring at the bottom of the pool into a boisterous and competitive team activity. He got involved in the debate club, an activity that teaches many high school students to appear confident even when they're not. He began to write more, first poetry, then some short stories. He has always been good-looking, with dark Mediterranean coloring and a nimble, flowing quality to his movements. Now he had his ear pierced and began dressing more like a Californian. "By the end of his junior year he was well known on campus and doing many different things," says van Hoy. "I don't know if he planned that or if it just happened, but every new thing he tried he enjoyed, and he ended up thriving in them all."

Oaz also became more sensitive to some of the stereotypes that surround mathematically talented students. At an academically high-powered school like Monta Vista, the traditional stereotype of young mathematicians — as socially inept, poorly dressed geeks with pocket protectors — does not have much resonance. The students have all heard stories of "quant jock" mathematicians who went to work for software and financial companies and became rich. But kids who are good at math still attract a measure of suspicion. Maybe in a place like Silicon Valley they're seen as having access to knowledge that can confer great power. Or maybe the old stereotypes have just been updated. The movie *Good Will Hunting* came out while Oaz was in high school. The young mathematician in the movie (played by Matt Damon, an obvious contradiction to the canard that all mathematicians are homely) is clearly very talented, though also troubled by a childhood of abuse and neglect. His mentor, a sophisticated and renowned mathematician now bereft of inspiration, takes an interest in the young man, mostly to further his own career. The message of the movie seems to be that people do mathematics to avoid the inevitable complications of life. Will forgoes a mathematics job to follow his girlfriend to California. And in the most obvious departure from reality, he's hardly ever seen doing math, despite the tremendous effort needed for anyone to become a proficient mathematician.

Oaz learned to disguise his interest in math when it served his purposes. "Math isn't the coolest thing in the world to do," he says. "I don't really go around telling people about it. It's not something that's always that socially acceptable. I think a lot of people who are good at math don't pursue it for that reason, and I think more girls than guys drop it for that reason."

But he continued to do math, and to do it extremely well.

After his junior year at Monta Vista, he qualified for the Olympiad team that traveled to Seoul. There he was awarded a gold medal, meaning that he finished in the top twelfth of the competitors. From the middle of the pack as a seventh-grader, Oaz had risen, four years later, to the top of the problem-solving world.

▽

Many social scientists devote sizable portions of their careers to tracking the lives of precocious children to see what smart kids do when they grow up. Yet when they have summarized their results, many have looked back on the years of interviews, questionnaires, and analysis with a twinge of disappointment. It's not that academically or artistically gifted children don't do well as adults; most do. But rarely do they meet the extravagant expectations of those who study them and who inevitably come to admire them.

The most famous longitudinal study of academically advanced children was begun in 1921 by Lewis Terman, a professor of psychology at Stanford University. Earlier in his career Terman had been the leading developer of the IQ test in the United States. In his study of what he called "child geniuses," Terman used nominations by teachers and parents, followed by IQ testing, to identify about 1,450 third- to seventh-grade boys and girls in California whom he deemed to have exceptional promise. By that he meant that their IQ scores were higher than about 135.

From the very beginning, the project had what are now recognized as major flaws. Because of his sampling technique, the "Termites," as they came to be known, were drawn from a relatively narrow section of society. Most were white, from the middle or upper class, and attending good schools. Terman strongly believed that intelligence resulted almost entirely from a person's genetic endowment (the first report of his project was entitled *Genetic Studies of Genius*). He thought that the bright children he was studying would become the great artists, writers, and leaders of the future. "Moderate ability can follow, or imitate, but genius must show the way," he wrote. He was a strong proponent of the eugenic nonsense fashionable in the first part of the twentieth century. He thought that intelligence testing would result "in curtailing the reproduction of feeblemindedness and in the elimination of an enormous amount of crime, pauperism, and industrial inefficiency." From our perspective the dangers inherent in his ideas seem obvious.

Terman administered an enormous battery of tests to his young subjects, including IQ tests, physical exams, and personality assessments. He interviewed their parents, teachers, and physicians. He collected data on their ancestors, the books they read, their nervous tics, the foods they ate, the hours they slept, and the cleanliness of their homes. He even pried into the most personal details of their upbringing and sexual lives as adults. For the most part the Termites endured this constant poking and prodding with stoic good humor, convinced that they were contributing to a worthy cause.

Terman expected his young subjects to become the leaders of their generation, as exceptional in adulthood as they were as children. In this respect, said Terman's successor at Stanford, Albert Hastorf, "It's my guess that Terman was a little bit disappointed." Only one of the 1,500 Termites became a well-known scientist — the physiologist Ancel Keys, who developed the portable meals now called K rations. In fact, two children who lived in California and would later go on to win Nobel Prizes — William Shockley, a coinventor of the transistor, and the physicist Luis Alvarez — were not included in the study because their IQ scores were too low. Probably the most influential of the Termites was Jess Oppenheimer, who became a humor writer and later created *I Love Lucy* and other well-known TV shows. Yet his success could not have been predicted. One of Terman's assistants wrote after an interview with him as a child, "I could detect no signs of a sense of humor."

Some of the Termites never seemed comfortable with the "genius" label (Terman himself gradually abandoned the term in favor of "gifted"). A few committed suicide, though no more than would be expected in any large sample of Americans, and several others developed severe psychological problems. Some floundered professionally, never quite deciding what they wanted to do.

But by far the majority of the Termites grew up to be well-adjusted, financially comfortable, conventionally successful professionals. They were engineers, doctors, lawyers, businessmen, and professors. Following the standards of the time, most of the women did not have careers, but they were busy and productive in their homes and communities.

Terman thought that the most intelligent of his subjects would become the most prominent, but that turned out not to be the case. As part of his study, the twenty-six Termites with IQ scores above 180 were compared with twenty-six random Termites with scores below that point. The high-IQ group scored no better on various measures of success than did the lower-IQ group.

But between those identified as more successful and less successful, differences did tend to emerge. The Termites who were most successful as adults tended to have high levels of energy, curiosity, and interest. They were more persistent and hard-working as children and had a broader range of interests. More of them had graduate degrees, and they earned their degrees at younger ages. They were more likely to say that they enjoyed their jobs.

That many of the Termites could not meet Terman's expectations is not surprising, say psychologists who have since conducted similar studies. As Ellen Winner writes in *Gifted Chil-*

dren, "Every prodigy eventually becomes an ex-prodigy." As academically advanced children grow up, they must make several critical transitions if they are to be comparably outstanding as adults. They have to convert the largely technical skills they have learned into a broader mastery of a field, so that they can begin to make original contributions to it. They must learn to rely on their own initiative and confidence rather than on the praise of parents and other adults. Many child prodigies undergo psychological crises during late adolescence when they realize that their skills, which were praised so extravagantly at younger ages, are no longer sufficient to excel. Some make the transition to new and deeper forms of accomplishment; others turn away from their early domain of expertise.

In mathematics, for example, skilled adolescents must move from being problem solvers to problem finders if they want to be professional mathematicians. A teenager can excel in school and in competitions like the Olympiad by becoming adept at solving problems for which an answer is already known to exist. But to become a research mathematician, a person has to be able to identify and make progress on interesting problems that may not have solutions.

Precocious children no longer have the field to themselves as they grow older, since many people who become prominent as adults did not exhibit particular signs of promise as children. In a study of 317 eminent people who lived during the twentieth century, the psychologist Victor Goertzel found that two-thirds were not described even as precocious when they were young, much less as prodigies. In an examination of the lives of Albert Einstein, T. S. Eliot, Sigmund Freud, Mohandas Gandhi, Martha Graham, Pablo Picasso, and Igor Stravinsky, Howard Gardner found that only Picasso could be considered a prodigy. Many people who go on to attain eminence do not discover the field in which they will excel until college or later. Others coast along in their chosen field until something makes them catch fire. And of course luck is an important factor; there's no substitute for being in the right place at the right time.

The word "success" is itself a loaded term. Many eminent people, though of course not all, must make difficult trade-offs in the pursuit of their goals, or trade-offs may be imposed on them. Some sacrifice close relationships with a spouse, children, and friends. Many prominent creators have had a stressful child-hood; a surprising number, for example, experienced the death of a parent while they were young. The occasional association of mental illness and creativity is another indication that some lives of great accomplishment are far from tranquil.

Besides, to judge the Termites as unsuccessful is to adopt Terman's own skewed perspective. By any conventional measure of success, Terman's subjects did extremely well. Most had good jobs, strong families, broad interests, and comfortable lives. That few became fame-obsessed painters, poets, or musicians is more a reflection of their overall good sense than any kind of lost promise. Those who lived into their eighties and nineties remained energetic, curious, and interested in the world around them.

▽

The 118 males — plus Melanie Wood — who have been on U.S. Olympiad teams constitute a much smaller sample than the 1,500 people in Terman's study. The Termites were Californians chosen largely on the basis of their IQ scores. The Olympians — from all over the United States — ran a much more demanding gauntlet. Each had to demonstrate the insight, perseverance, talent, and creativity needed to be numbered among the best problem solvers in the world. They rightly consider themselves members of a small and very select group.

A logical temptation is to think that all the Olympians should be as exceptional in adulthood as they were as high school mathematicians. As the Termites' experiences demonstrate, that expectation is obviously unrealistic. Many Olympians have fairly typical jobs in academia, business, or government. But a closer inspection reveals many ways in which they have begun to distinguish themselves.

The best example is Eric Lander. After competing with the first U.S. Olympiad team in 1974, he studied mathematics in college, earning an undergraduate degree from Princeton and a Ph.D. from Oxford in 1981. He became interested in biology while teaching at Harvard Business School in the 1980s, partly through conversations with a brother who was a neuroscientist. He taught himself molecular biology, became a faculty member at MIT's Whitehead Institute, and in 1990 founded the institute's Center for Genome Research, which quickly became a leader in the effort to sequence the human genome. "Eric Lander came through in the final stretch," writes Ingrid Wickelgren in her book *The Gene Masters*, "applying his exceptional talents for automating biology to sequencing and churning out the most human draft sequence of any laboratory in the public effort." The politics of the Nobel Prize are complicated; recipients have to be honored while they are alive, so a backlog of aging notables always exists, and each award can go to no more than three individuals, which means that team projects are sometimes slighted. But Lander, a man who has already made contributions to science that will be remembered for centuries, is certainly on the Nobel short list.

Other Olympians have been gaining renown as they have entered their prime professional years. Peter Shor, a member of the 1977 team that finished first in Yugoslavia and now a mathematician at AT&T Labs in New Jersey, has been a leader in the development of quantum computing — an effort to use the quantum properties of atoms to produce computers that are orders of magnitude more powerful than today's. Several other prominent young professional mathematicians in the United

States were on Olympiad teams. Two Olympians founded software companies, and one remains the chief engineer of his company. Two are Talmudic scholars. One performed with an ensemble at Carnegie Hall.

Much of this information comes from James Campbell, who leads a research project at St. John's University in New York City that is tracking the careers of the Olympians. For twenty-five years Campbell ran the Metropolitan New York division of the Junior Science and Humanities Symposium, a regional and national competition for high school students who have done original research in science, engineering, and mathematics. He decided to try to find out whether competitions made a difference in students' lives.

Like some of the Termites, a few Olympians had problems in college, and others searched long and hard for a challenge that would match their earlier triumphs. But those are the exceptions, Campbell says. Most former team members went to prestigious colleges, with Harvard and Princeton the two top choices. More than half have earned doctoral degrees (the comparable figure for Terman's study was about a quarter of the eight hundred males). Those old enough to have now finished college and graduate school generally have good, and in many cases high-paying, jobs.

Campbell's most remarkable finding is how wide-ranging the Olympians' career choices have been. Only about a quarter of them have become research mathematicians at universities. Some of the jobs held by others are still mathematically intense — such as those held by the several who work on top-secret national security projects. But other Olympians don't do much math anymore. Some have become lawyers, engineers, and doctors. Others have applied their talents to more lucrative ends.

Take Eric Wepsic, who now works for the D. E. Shaw group, a securities trading and investment firm in midtown Manhattan. Wepsic was on the 1987 and 1988 teams that competed in Cuba and Australia. After receiving his baccalaureate from Harvard, he began graduate school in mathematics there, but he was "unhappy with research," he says. "I had never decided on grad school. At the time I saw it more or less as my only option, which is a really stupid way to choose a career."

Wanting to do something faster-paced and more engaging, Wepsic got in touch with a friend who had gone to work for Shaw, applied for a job, and was quickly hired. At first he did various kinds of research connected with equities. Now he manages most of the automated trading groups at Shaw. He says that he still uses a lot of math in his work, "like probability and statistics, and also a lot of mathematical reasoning, which we apply to our trading algorithms." And even though he doesn't use much of the math he learned after his first couple of years of college, "the mental training that I got in math has been very helpful."

Wepsic encourages young mathematicians to "try to vary what you do (at least in small ways) so that you continue to learn." Most of the Olympians don't need much encouraging to be venturesome. They are curious about almost everything — music, games, sports, literature. A surprising number have written short stories or poetry, and one star Olympian during the 1980s, Princeton math professor Jordan Ellenberg, has written a delightful and widely acclaimed novel, *The Grasshopper King*. And when they try something new, they typically learn to do it well.

Being good at so many things has only one major drawback, say the Olympians and others who know them. When mathematicians already susceptible to arrogance decide they can do anything they set their minds to, their self-regard can become a bit overwhelming. The questionnaires Campbell sends to the Olympians inevitably contain a few typos, and just as inevitably a few are returned with the typos corrected. "It doesn't bother me, because I'm used to it," Campbell says. "But I can imagine how a boss would feel."

▽

Oaz drew on a particular kind of breadth in solving problem five — an ability to pursue the obvious relentlessly. Like problem one, it was a geometry problem. And also like problem one, it sounded more complicated than it really was.

> Let ABC be a triangle with angle BAC = 60 degrees. Let AP bisect angle BAC and let BQ bisect angle ABC, with P on BC and Q on AC. If AB + BP = AQ + QB, what are the angles of the triangle?

The first thing Oaz had to do was convert this complicated-sounding problem into a diagram. So he drew the following sketch:

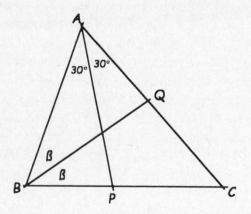

The problem states that the angle BAC is 60 degrees and that line AP bisects angle BAC. Oaz therefore knew that each of the two halves of the angle equaled 30 degrees, which he indicated on his diagram. Similarly, the problem says that line BQ bisects angle ABC. But Oaz did not know what angle ABC was, so he labeled each half of the angle with the Greek letter beta (β). If he could figure out what angle beta was, then he would know that angle ABC was twice that, and that would give him the solution to the problem.

This problem has several extremely clever solutions (one of which is described in the appendix). The trouble with such solutions is that you can spend a lot of time looking for them, and if your cleverness fails you, a lot of time can be wasted. Oaz didn't want to take a chance with problem five. He decided to dumb-ass it.

Dumb-assing, in Olympiad parlance, is what you're doing when you're not being clever. You take the most obvious approaches and try to work through a problem by brute force. "To some extent, dumb-assing is a state of mind," says former Olympian Alex Saltman. "Dumb-assing gives you something to do that could give you inspiration, or maybe even provide you with a complete proof. What you try to do is minimize the number of ugly steps in a dumb-ass proof so they're not all ugly."

Oaz began by defining the lengths of various line segments, like AB and BP, in terms of the sines and cosines of 30 degrees and angle beta. He then converted the statement AB + BP = AQ + QB into its trigonometric equivalent and began manipulating it. At times the equations got pretty hairy. A key step in his solution was the equation $\sin [60 - (\beta/2)] = 2 \sin (30 + 2\beta) \sin (\beta/2)$. Eventually Oaz ended up with two equivalent expressions that could be plotted as lines on a plane. The two lines intersected at a point indicating that beta had to be 40 degrees. Therefore, angle ABC in Oaz's diagram had to be 80 degrees. Since the angles of a triangle always add up to 180 degrees, and the problem says that angle BAC is 60 degrees, angle ACB had to equal 40 degrees.

Oaz's solution for problem five was far from elegant, but it worked; his solution received a perfect score. It was part of a superb second day for Oaz, who received perfect scores on problems four and five and partial credit on problem six. Oaz's resourcefulness as a problem solver had come through again.

Yet this Olympiad had a wistful feeling for Oaz. The goal he had sought for so long — to be numbered among the elite of high school mathematicians — had now been achieved. "I like math, for fun and for competitions," he said. "But I don't want to do it forever. If you do research and become a professor, the math that you do can be really abstract. I don't really want to do that. I might do something that's related to math, like some kind of science. But I might also do something completely unrelated."

9. a sense of wonder

The pressure on the competitors seemed greatest during the evenings right before the Olympiad, so the organizers decided to fill one of the evenings simply with a movie. The Olympians speculated about what the movie might be — a James Bond flick, maybe, to take their minds off mathematics? But then word got around: "It's that *Nova* show where Andrew Wiles cries." Many of the competitors had already seen it. Yet almost all of them were quietly watching as the lights dimmed in the viewing room of the student union building.

The show opens with the moment the Olympians remembered, which occurred during an interview with Wiles in the attic of his home in Princeton. There, after years of work, he finally had proved Fermat's last theorem. "At the beginning of September," says Wiles, "I was sitting here at this desk, when suddenly, totally unexpectedly, I had this incredible revelation. It was the most — the most important moment of my working life. Nothing I ever do again will — I'm sorry."

Wiles's proof has been called one of the greatest triumphs of twentieth-century mathematics, yet it began with a set of observations no different from those in an Olympiad problem. According to the Pythagorean theorem, if the sides of a right triangle are labeled x, y, and z, with z the longest side, then the lengths are related by the formula $x^2 + y^2 = z^2$. For example, a right triangle with sides of lengths 3, 4, and 5 satisfies the Pythagorean theorem because $3^2 + 4^2 = 5^2$, or $9 + 16 = 25$. The Greeks knew that an infinite number of different triangles with whole-number sides satisfy the equation. Thus, an infinite number of whole-number triads (like 3, 4, and 5; 7, 24, and 25; 8, 15, and 17; and so on) must exist that solve Pythagoras's equation.

Sometime in the 1630s the French mathematician Pierre de Fermat — the same Fermat who occasionally challenged his colleagues with difficult problems — was reading a Latin translation of a book called *Arithmetica* by the Greek mathematician Diophantus of Alexandria. One portion of the book was devoted to observations about the Pythagorean theorem, and Fermat, as he read, began wondering about possible extensions of the theorem. For example, could he find a trio of whole numbers that solved $x^3 + y^3 = z^3$? More generally, could he find trios of whole numbers that would satisfy $x^n + y^n = z^n$, in which n is any whole number greater than 2?

In the margin of his copy of the *Arithmetica*, Fermat wrote a note that has haunted mathematicians ever since. He claimed that no whole-number solutions exist for $x^n + y^n = z^n$ if n is a whole number greater than 2. In other words, for $n = 3$ and for all other values up to infinity, three whole numbers would never be found that satisfy the equation. Then he wrote, "I have a truly marvelous demonstration of this proposition which this margin is too narrow to contain."

Fermat's note came to light only after his death many years later, at which point mathematicians began looking for his "demonstration." But exhaustive searches of his mathematical papers, his library, and even his home did not turn it up. Meanwhile, the world's best mathematicians sought to prove his conjecture — without success. Over time the many other mathematical propositions that Fermat had claimed without proof were solved. But his claim that $x^n + y^n = z^n$ has no whole-number solutions when n is greater than 2 resisted their best efforts. The statement therefore became known as Fermat's last theorem — even though it hadn't been proved and was thus a conjecture rather than a theorem, and even though Fermat was a relatively young man when he proposed it.

Andrew Wiles became interested in Fermat's last theorem when he was ten years old. At a library in Cambridge, England, near his childhood home, he picked up a book called *The Last Problem* by Eric Temple Bell, which told the story of Fermat's conjecture and explained how it had defied mathematicians for more than three centuries. "It looked so simple, and yet all the great mathematicians in history couldn't solve it," Wiles told Simon Singh, the producer of the show the Olympians watched that evening. "Here was a problem that I, a ten-year-old, could understand, and I knew from that moment that I would never let it go. I had to solve it."

Wiles worked on the problem for several years when he was young but progressed no further than any other mathematician had. As a teenager he became so interested in mathematics that he decided to make it his career, and he stopped working on Fermat's last theorem as he came to appreciate its difficulties. But the problem was always in the back of his mind. After receiving his doctorate in mathematics from Clare College, Cambridge, in 1980, he moved across the Atlantic to Princeton University, where he became a rising star in the mathematical field known as number theory. Then, in the mid-1980s, new developments in mathematics revealed a possible new approach to Fermat's last theorem. Setting aside his other interests, Wiles devoted seven straight years of his life to his childhood obsession, working largely in seclusion in the attic of his Princeton home. Finally, on the morning of September 19, 1994, he realized that he had done it — he had solved a problem that had stymied mathematicians for hundreds of years.

Wiles's inability to describe his breakthrough without choking up may seem slightly absurd. How could someone get so emotional about a math problem? Yet the Olympians watched him relive that moment with quiet respect. Each of them seemed to know exactly how he felt.

$\triangledown$

The sixth and last problem on the Forty-second Olympiad — by tradition the hardest of all — looked deceptively straightforward to the competitors.

> Let $a > b > c > d$ be positive integers and suppose that $ac + bd = (b + d + a - c)(b + d - a + c)$. Prove that $ab + cd$ is not prime.

The problem comes from the branch of mathematics known as number theory, which investigates the properties and relationships of whole numbers. It asks you to prove that if you have four whole numbers of decreasing values that meet a particular condition, then the first two numbers multiplied together plus the second two numbers multiplied together cannot be a prime number. In other words, $ab + cd$ has to be evenly divisible by at least one whole number other than 1 and itself. For example, 21 is not prime, because it is the product of 3 and 7.

Many hours of trial and effort can produce four numbers that satisfy the condition laid out in problem six. For example, the numbers 21, 18, 14, and 1 do the job, because $(21 \times 14) + (18 \times 1) = (18 + 1 + 21 - 14) \times (18 + 1 - 21 + 14)$. And sure enough, the number $(21 \times 18) + (14 \times 1) = 392$ is not prime, because $392 = 2 \times 2 \times 2 \times 7 \times 7$. But the Olympians had to do much more than find a single example that satisfies the problem. They had to prove, using the methods of number theory, that the statement holds for any four whole numbers, from zero to infinity.

Number theory sometimes has been called the "queen of mathematics," though the phrase is not necessarily complimentary. Historically, number theory was seen as elegant but somewhat irrelevant to the real work of mathematics, a pleasant diversion that had to be put aside when the heavy lifting of solving a scientific problem began. The number theorist G. H. Hardy (who also called mathematics a "young man's game") wrote in his 1940 autobiography *A Mathematician's Apology*, "I have never done anything 'useful.' No discovery of mine has made, or is likely to make, directly or indirectly, for good or ill, the least difference to the amenity of the world."

As with his generalizations about age, Hardy sold number theory short. The codes that keep nuclear weapons from being launched without authorization are derived from number theory, as are a host of other encryption techniques. Applications of number theory turn up in physics (calculating atomic energy levels), acoustics (designing sound diffusers), and information theory (devising error-correcting codes). Hardy himself once wrote a letter to *Science* magazine describing the transmission of genetic traits in populations, and the resulting formulation — though technically not a result of number theory — is known today as the Hardy-Weinberg equilibrium.

But perhaps the greatest influence of number theory has been its appeal to the imagination. Countless children have spent hours and days and years wondering why numbers work the way they do. Why is it that every even number greater than two that has ever been tested is the sum of two prime numbers? (The proposition that *all* even numbers meet this condition, known as Goldbach's conjecture, has never been proved.) What determines the spacing of prime numbers along the number line? (The distribution of primes is the subject of another famous unsolved conjecture called the Riemann hypothesis.)

Number theory was one of the subjects — along with logic and algebra — that captured the childhood attention of Gabriel Carroll, the sixth member of the U.S. Olympiad team. When he was very young he began reading simple math books and working his way through logic puzzles. "I definitely got an early start," he says. "I remember I was doing arithmetic and algebra when I was six." Soon he was reading books about recreational mathematics, especially those by Martin Gardner, and working on more advanced problems. "I would randomly come up with theorems and prove them on my own — easy stuff. I remember the first thing I proved that I could call a theorem — I was about ten years old at the time — was that the perpendicular bisectors of the sides of a triangle come together at a point. So I went home and told that to my mom, and she said, 'Yeah, that's a well-known result.' I was really disappointed."

Gabriel was the only one of the six U.S. team members who had not participated in Mathcounts. He grew up in Oakland and attended the public schools in the city. When he was in the fifth grade he took part in a local mathematics competition that was so unsatisfying that he lost his enthusiasm for contests. "When I was offered the opportunity to compete in Mathcounts in the seventh grade," he says, "I was given the impression that it was another competition like the previous one, so I decided not to do it." But he continued to read math books on his own. In the eighth grade he took the American High School Mathematics Exam and did so well that he was asked to take the American Invitational Mathematics Exam and the U.S. American Mathematical Olympiad — a remarkable achievement for an eighth-grader. His success rekindled his interest in competitions, and he began to work harder on problem solving.

The next year he went to Oakland Technical High School, a large public school that is divided into quasi-vocational programs called academies; Gabriel was in the engineering academy, which he describes as a combination of drafting and physics. Gabriel took calculus in his freshman year at the recommendation of a teacher who noticed his abilities. At the end of that year he qualified not only for the Olympiad summer training program but also for the Olympiad team. "I worked with him first at the summer camp," says the Berkeley mathematician Zvezdelina Stankova, who is an assistant coach of the Olympiad team. "Gabe was not really known before that, at least not nationally, and he had not been to the summer camp before. So his qualification for the team was a big surprise. We weren't sure what to expect from him, and whether he would be up to the other members of the team. But Gabe joined the lectures as if he had been there for years. He shone from the very beginning. Often he not only solved a particular problem but also offered alternative ways to solve the problem or even generalizations of the problem. I was very impressed by him."

At the Olympiad in Taiwan that year, Gabriel earned a gold medal, as did his teammate Reid Barton, also a freshman. In his next year at Oakland Technical High, Gabriel began taking math classes at Berkeley, a fifteen-minute bus ride from his school. He also started attending the math circles that Stankova and other mathematicians in the Bay Area were organizing. "He learned a lot there, and also met a lot of other people interested in mathematics," says Stankova. "The circles are not just about problem solving. Some lectures are all mathematical theory with no direct applications to Olympiad problems, so they enlarge your understanding of what is going on."

If anyone on the Olympiad team looked like a Californian, it was Gabriel, with his dirty-blond hair, wire-rim glasses, and a bit of a stubble. He has a loose-limbed, lanky stance, as if his body were suspended from his shoulders. He is a master of mirror writing. If you say a sentence, he can immediately write it backward and then say it backward. "It's just something I picked up along the way," he says.

Sometimes Gabriel has a distracted air, as if he isn't paying much attention to what's going on around him. But that's a ruse. On his personal Web site, built around an elaborate persona called the Gastropod, he has compiled hundreds of snippets of conversation gathered over the years. Some are Yogi Berraisms: "Anyone who remembers this is not ever gonna forget it." "Conversation is so boring when everyone is introverted" (attributed to Oaz Nir). Others are mathematical: "This is a really good proof. It's been around for 2300 years, but it still works." "I set it to circumcise the hexagon around this circle." And others are odd amalgams of mathematics and life: "I will give you two answers to that, one smart and one wise." "That annoying e^{-rt} [a term that gets smaller and smaller with time] is because you don't live forever. Because eventually you die."

▽

Gabriel's answer to problem six demonstrated his power as a mathematician. "Gabe's solution was overkill," says Stankova, "but he solved the problem the way a mathematician would solve it." In his solution he used a mathematical idea called a ring — a set of mathematical objects, any two of which can be added or multiplied to yield another member of the set. For example, the whole numbers constitute a ring, because adding or multiplying two whole numbers produces another whole number. But rings can be constructed using other kinds of mathematical objects, and Gabriel, in solving problem six, used one of the most exotic categories of elementary mathematical objects to build his ring — imaginary numbers.

Many people give up on math just about the time they learn about imaginary numbers. That's unfortunate, because imaginary numbers are fascinating and not that complicated. They involve square roots, which for positive numbers are straightforward: the square root of 4 is 2, because $2 \times 2 = 4$. But what is the square root of -4? Multiplying a negative number by a negative number always gives a positive number. Similarly, multiplying a positive number by a positive number always produces a positive number. So what number can be multiplied by itself to yield a negative number?

As it turns out, none of the counting numbers, those we learn about in grade school, work. But mathematicians have learned over the centuries that the square root of a negative number is just as legitimate a number as any counting number. The difference is that we cannot interpret its physical reality in the same way as for a number like 2.5, which can be seen as a point on the number line. That's why mathematicians began calling the square roots of negative numbers imaginary numbers, in contrast to the "real" numbers on a number line. And they gradually adopted a special symbol, i, for the simplest imaginary number, the square root of -1.

In his solution to problem six, Gabriel used a number known as omega, which is defined in terms of i, to factor the equation (omega is described in more detail in the appendix). Then he was able to show that $ab + cd$ is the product of two whole numbers, which he called p and r. To prove that $ab + cd$ is not prime, as the problem required, he had to show that neither p nor r could be 1. He assumed the opposite and solved the problem by showing that the assumption led to a contradiction.

The power of Gabriel's proof derives in large measure from the capabilities mathematicians gained when they learned to use imaginary numbers in their work. Many other examples of these capabilities can be cited. Imaginary numbers are used to analyze electrical currents, the motions of springs, the flow of fluids, and countless equations involving continuous change. They come up in signal processing, number theory, and quantum mechanics.

Imaginary numbers also figure in one of the most famous equations in all of mathematics:

$$e^{\pi i} = -1$$

In its six symbols this equation unexpectedly unites four areas of mathematics that seem almost completely distinct: arithmetic, analysis (a branch of mathematics based on calculus), geometry, and algebra. The number -1 is familiar to everyone from the arithmetic learned in school. The number e, which comes up often in calculus, equals $2.71828\ldots$, with the ellipses indicating that the string of digits giving its exact value extends to infinity. The more familiar number π — $3.14159\ldots$ — is the ratio of a circle's circumference to its diameter. And i is the square root of -1. Thus if e is raised to the πi power, the result is -1. The equation was first derived by the great eighteenth-century Swiss mathematician Leonhard Euler, and it has fascinated mathematicians and scientists ever since (the physicist Richard Feynman once called it "the most remarkable formula in mathematics"). As the nineteenth-century Harvard mathematician Benjamin Peirce said: "It is absolutely paradoxical; we cannot understand it, and we don't know what it means. But we have proved it, and therefore we know it must be truth." Some mathematicians have considered this equation so transcendent that they have had it engraved on their tombstones.

▽

Unexpected connections among different branches of mathematics have been a prominent theme in the history of the discipline. René Descartes united algebra and geometry by showing that shapes could be expressed as equations. The fundamental theorem of calculus derived from the work of Isaac Newton and Gottfried Leibniz related the boundaries of geometric objects to properties of their interiors. Georg Cantor demonstrated the existence of different kinds of infinities by examining bounded intervals on a number line. When students grasp these profound insights for the first time, the frustrations and difficulties of learning mathematics can fall away. They sense a kind of intelligence at work beneath the buzzing confusion of everyday life. The world makes sense: it can be described mathematically.

Mathematical connections were the centerpiece of Wiles's work on Fermat's last theorem. By the mid-1980s only amateurs and crackpots were spending much time on the problem. "Fermat's last theorem was viewed as a curiosity," says Ken Ribet, a mathematician at Berkeley whose office is just upstairs from Stankova's. "From the viewpoint of modern mathematics, it had no special interest aside from its long history. It was impossible to say why a person should be interested in that equation as opposed to any other. It didn't seem to have any connection to anything else."

Then an unexpected mathematical connection suddenly brought the theorem back into play. In 1955 the Japanese mathematician Yutaka Taniyama proposed that a deep and mysterious link existed between two distinct areas of modern mathematics. One area involves elliptic curves, which are mathematical objects described by a particular kind of equation. The other area involves modular forms, which are sets of functions that exhibit certain symmetric properties. No one before Taniyama would have guessed that elliptic curves are necessarily related to modular forms. But he proposed that the two subjects were flip sides of the same coin. He said that every rational elliptic curve with rational coefficients has a corresponding modular form, and vice versa.

This conjecture, which was made more precise by Taniyama's friend and colleague Goro Shimura, shocked mathematicians. It meant that someone working on a problem involving elliptic curves could instead investigate a comparable problem involving modular forms. A problem that was very difficult in one field could be translated into the other, where a solution might be much easier. Mathematicians quickly realized that the Taniyama-Shimura conjecture was an extremely powerful mathematical tool. They therefore began to use it even though it had not been proved. That is, they would begin a proof by saying, "Assume that the Taniyama-Shimura conjecture is true. In that case, the following very interesting results can be demonstrated."

But there was a problem. Mathematicians soon realized that proving the Taniyama-Shimura conjecture would be extremely difficult. It would require new mathematical ideas and tools that were unlikely to be developed anytime soon. And without a proof of the conjecture, any results built on it would inevitably be suspect.

In 1985 the German mathematician Gerhard Frey dropped another bombshell, pointing out that if the Taniyama-Shimura conjecture could be proved, it also would imply Fermat's last theorem. Here's why. If the equation $x^n + y^n = z^n$ in fact did have a whole-number solution for n greater than 2, that solution could be converted into a bizarre elliptic curve that almost certainly did not have a corresponding modular form. So Fermat's last theorem could be proved by contradiction. If every elliptic curve with rational coefficients has a modular form, as Taniyama proposed, then a set of whole numbers satisfying $x^n + y^n = z^n$ could not exist and Fermat's last theorem would have to be true.

Frey came up with the idea, but he could not conclusively prove the connection between the Taniyama-Shimura conjecture and Fermat's last theorem. That task was accomplished by Ken Ribet, who wrote his proof in 1986–87. When Wiles heard about Ribet's proof, he immediately realized that his childhood dream was again within reach. If he could prove the Taniyama-Shimura conjecture, he would automatically have proven Fermat's last theorem.

By this time Wiles was a well-known mathematician. Now he realized that if he wanted to prove Fermat's last theorem, he had to make a break with his professional past. He decided not to tell his colleagues at Princeton that he was working on the Taniyama-Shimura conjecture. Most of them would have thought the effort futile, and he didn't want to be discouraged by their pessimism. Undoubtedly he also had his eyes on the history books. He knew that whoever solved Fermat's last theorem would be remembered for as long as the world has mathematicians. He didn't want to do most of the work on a proof only to have someone swoop in at the last moment and steal it away from him.

Throughout the latter part of the 1980s and the early 1990s, Wiles continued to work on his proof and teach classes at Princeton. But he quit attending most mathematics meetings. Occasionally he published a paper on a minor topic to show that he was still doing research. Many of his colleagues probably figured he had burned out and would no longer make any important contributions to mathematics.

In early 1993 Wiles thought that he had succeeded. He had combined a wide range of extremely advanced mathematical techniques to prove a particular part of the Taniyama-Shimura conjecture, which was enough to prove Fermat's last theorem. He finally decided that he was ready to tell the world about his work. At a mathematical meeting in Cambridge, England, Wiles sprang the news on his unsuspecting colleagues. The announcement generated worldwide headlines. He was featured in *People* magazine, interviewed on CNN, and even asked to advertise a line of clothes.

But before he could publish his proof in a mathematical journal, it had to be checked by other mathematicians. The manuscript of his proof was sent to six number-theory experts, who promised not to reveal the contents of the proof until it was published. Within a few weeks the reviewers uncovered a flaw: one of the techniques Wiles had developed to prove the conjecture had a gap. Unless the gap could be filled, the proof would fail.

Now the stakes were extremely high. Wiles had revealed his dreams to the world, yet he appeared to have fallen short. He worked month after month on the gap but made little progress. In the show the Olympians watched, Wiles said, "After there was a problem with [the proof], there were dozens, hundreds, thousands of people who wanted to distract me. Doing mathematics in that kind of rather overexposed way is certainly not my style, and I didn't at all enjoy this very public way of doing it."

By early 1994 Wiles was in despair. He saw no way to fill the gap in his proof, and other mathematicians were clamoring for him to publish the details of everything he had done up to that point so they could get to work on the problem. Finally Wiles decided that he needed help, so he asked Richard Taylor, a young mathematician at Cambridge University, to work on the proof with him. Even together, the two made little progress. In his book *Fermat's Enigma*, Simon Singh has a particularly evocative description of this period. "Having ventured farther than ever before and failing over and over again, they both realized that they were in the heart of an unimaginably vast labyrinth. Their deepest fear was that the labyrinth was infinite and without exit, and that they would be doomed to wander aimlessly and endlessly."

Toward the end of that summer, Wiles was ready to tell Taylor that they should admit defeat. More than a year had passed since his Cambridge announcement, and he and Taylor were making no progress on the gap. Maybe he should just publish the flawed proof so that someone else could figure out how to fix it. But Taylor argued that they should work together for another month, until the end of September, and then move on.

Wiles decided to use part of that time to examine the method he had used in the original proof. Why had the method turned out to be flawed? Suddenly, on that fateful morning in September, he realized how to fix the problem. He could draw on an earlier method he had abandoned as insufficient. Neither method alone was enough to do the job, but the two complemented each other perfectly. Immediately Wiles knew that the proof was complete. Every elliptic curve he was studying had a corresponding modular form. He had proved Fermat's last theorem.

▽

The Taniyama-Shimura conjecture is part of a much larger, almost metaphysical effort to unify much of modern mathematics. In the 1960s Robert Langlands of the Institute for Advanced Study in Princeton proposed that many areas of mathematics are linked in the same way that elliptic curves and modular forms are linked. These linkages, if they can be demonstrated, would have wide-ranging implications. As with the Taniyama-Shimura conjecture, they would enable mathematicians to move a problem from one mathematical domain to another, where it might be more easily solved. More generally, these linkages would demonstrate a profound interconnectedness in mathematics, as if something (some people might say someone) was making the pieces fit together.

"The various branches of mathematics are all interconnected," says Berkeley's Ribet. "Mathematicians work on very narrow problems. They bang their heads against the wall, trying to answer some small question. But when they step back and look at what people are doing in other areas of mathematics, they often find that people are using similar tools, similar methods, similar philosophies. . . . The main thing that mathematicians try to do is to understand some situation, which basically means bringing order to it. When the understanding involves a bridge of some sort between parts of mathematics that are generally thought to be far from each other, this is always an exciting and unexpected development."

No one understands exactly why the various branches of mathematics are linked. Perhaps the links reflect a deeper underlying logic that mathematicians have only begun to grasp. Perhaps mathematics will someday be conducted on a different level that will subsume all of today's mathematics. But how would that explain another striking aspect of modern research: the emergence of deep and fundamental connections between mathematics and the functioning of the natural world? "Miraculous connections have been emerging," says Harvard professor Arthur Jaffe. "Ideas from theoretic physics are providing bridges to different areas of mathematics. These bridges provide insights so far-ranging that we don't understand where they come from."

At the time of the Olympiad, Jaffe was president of the Clay Mathematics Institute, which was established by the mutual funds magnate Landon Clay to promote mathematical research. At the beginning of the year 2000, the institute announced that it would award prizes of $1 million each for solutions to seven "Millennium Problems" — problems so difficult that they have remained unsolved for decades or centuries. All seven problems feature connections among different areas of mathematics and among mathematics, science, and engineering. The Riemann hypothesis has connections to encryption techniques and to an area of physics known as quantum chaos. Yang-Mills theory relates the mathematics used to study fundamental particles with the mass of those particles. The P versus NP problem indicates whether certain computations can be performed with a computer. Solutions to the Navier-Stokes equations would help engineers design objects that flow more efficiently through fluids.

The Millennium Problems are perfect examples of math's connectedness and mystery. Mathematicians working on their own little tasks often stumble across a connection to one of the Millennium Problems, as if they had followed a random path through the woods that opened onto the end of a rainbow. "Mathematics is a big whole," Jaffe says. "If you're original, creative, and broad-minded, eventually you'll end up in every subfield of mathematics, and the work you're doing will have an influence throughout mathematics. It makes it hard to be a mathematician, because you have to be a student for your entire life."

▽

Many mathematicians have two sides to their personality. The objects that they manipulate in their minds are the products of reason and logic. Yet mathematicians are often intensely emotional people. Many fall in love with the subject and never lose their passion for it. "I don't do mathematics because it's important," Gabriel once told an interviewer. "I do it for aesthetic reasons. Math is an art."

Andrew Wiles would agree. At the moment when he achieved his breakthrough, he did not reflect on the acclaim his work would bring. He was struck by the beauty of what he had done. "It was the most — the most important moment of my working life. Nothing I ever do again will — I'm sorry. It was so indescribably beautiful; it was so simple and so elegant. I just stared in disbelief."

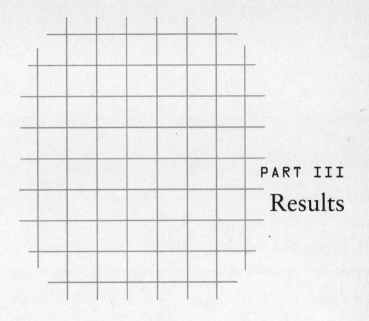

PART III

Results

10 · triumph

The health club at the Embassy Suites hotel in Washington, D.C., is separated from the second-floor corridor by a wall of glass. As people staying at the hotel walk down the corridor, they can't help but glance through the window at the sweaty exercisers perched on their Lifecycles and StairMasters on the other side. The same incongruity applies inside the health club; as its patrons are straining at the Cybex machines, a steady stream of hotel guests — showered, well dressed, hurrying to their destinations — is forever passing by.

Two days after the conclusion of the exam at the Forty-second Olympiad, a quite different spectacle began to take shape outside the health club's glass wall. Every few minutes a disheveled, worried-looking man appeared and added a piece of paper to the other papers taped to the glass. Immediately a dozen other disheveled, worried-looking men clustered around and studied the paper intently, some taking notes on pads in their hands. A member of the health club who was skilled at mirror writing could peer through the papers and see that each was headed by the name of a country — Ecuador, Israel, China. Other than that the papers bore only meaningless columns of numbers.

To the coaches on the other side of the glass, those numbers were anything but meaningless. They were the scores each team member had achieved on the six Olympiad problems. These scores would determine which Olympians received gold, silver, and bronze medals; which countries had reasons to celebrate or despair; which team could claim bragging rights as the strongest group of young mathematicians in the world. They were judgments from on high, as stark and unforgiving as the solid black type in which they were written.

∇

Because mathematics at this level is as much an aesthetic as an intellectual endeavor, the judging of the Olympiad is both objective and subjective — more like judging a figure skating competition than like grading a multiple choice test. The scores for each problem can range from 0, for no substantive progress made, to 7, for a correct solution. The scores from 1 to 6 are the tricky ones. The contestants may receive 1 or 2 points if they made some progress on a problem but remained far from a solution, 3 or 4 points if they got about halfway there, or 5 or 6 points if they were just a few steps from the answer. Where on that scale each Olympian's solution falls is a matter of judgment.

The scores are decided by panels of mathematicians known as coordinators. The coaches and assistant coaches appear before the coordinators and discuss each of their team members' solutions. The coaches and coordinators then try to agree on a fair score. "I think of this as a sort of legal proceeding," says Kiran Kedlaya, the assistant chief coordinator at the Forty-second Olympiad and a member of the 1990, 1991, and 1992 U.S. teams. "The student is the client, and the coordinators are the judge and jury. The coaches are the students' lawyers, and they argue the case."

Coaches take different approaches to the process, from mild to fiery. "Some coaches let the papers speak for themselves," says Kedlaya. "The leader from China is famous for this. He doesn't engage in much advocacy. He just says what's there. Other coaches try to present the solutions as favorably as possible. Some are a lot more aggressive than others. You might even use the word 'obstinate' in some cases."

On the spectrum from easygoing to obstinate, Titu

Andreescu definitely leaned toward the latter. He believed that his job was to make sure his students received every point they deserved to receive. After the exam he and the assistant coaches grilled the team members to figure out how to squeeze every possible point from an answer. "He's not the most aggressive," says Kedlaya. "But no one ever accused him of not presenting his students' papers in the best possible light."

In the early stages of the coordination, Titu's strategy appeared to be paying off. The scores for the teams tend to be reported in order, from the first problem to the last, though appeals and other controversies invariably confuse the process. On problem one the U.S. team got five 7s and just one 6. That surprised the team members. All had solved the problem, but several thought that slight errors in the logic of their proofs would cost them points. Titu and the other U.S. coaches must have been able to convince the coordinators that the errors were not serious enough to warrant subtractions.

But other teams also were doing well. The Koreans' boasting at lunch after the first day seemed justified. They scored straight 7s on problems one and two, whereas the U.S. scores fell off slightly on problem two. The Chinese team also appeared to be doing extremely well, with one 4 and eleven 7s on problems one and two.

But on the much harder problem three, the Koreans faltered, getting just four 1s, a 2, and a 3. The United States, in contrast, got perfect scores for Reid's and Gabriel's solutions and partial credit for the others. It began to seem that the U.S. team might emerge with the highest team score of all for the first day, and if they did that, they knew they had a chance to win the competition. But then the Chinese scores for problem three came up — *three* 7s, a 2, and two 0s. Of the 126 points possible on the first day, the Chinese had scored 104, compared to the Americans' 100.

▽

The prowess of the Chinese team — at the time of the Forty-second Olympiad, China had finished with the top score in three of the previous four years — is a popular topic of speculation at the Olympiad. The size of China's population is one often-heard explanation. If outstanding mathematical talent arises randomly at very low frequencies, then countries with the largest populations should have the largest number of skilled mathematicians. But this formula doesn't work very well for Olympiads. Brazil, the world's fifth-largest country in terms of population, generally does well at the Olympiad, but its team is not usually among the top finishers. In contrast, small countries with strong mathematical traditions, like Bulgaria and Korea, often do much better than their size would suggest.

A much more important factor is an educational system that gets children interested in math and then identifies those who are doing well. By these measures China excels. Mathematics teachers in China do not necessarily have more training than U.S. teachers. But they tend to be specialists, teaching only mathematics, and they study their craft hard. U.S. elementary school teachers, in contrast, typically teach all subjects — and math is a subject that makes many of them uncomfortable.

In the United States the usual way for children in middle school to become interested in math competitions is through Mathcounts. But if teachers don't know about Mathcounts and don't build on a student's interest in math, many potentially strong competitors never get started.

In China, far more teachers are interested in math competitions and are able to do the math associated with them. In essence China has a network of coaches throughout the country who identify and help train mathematically talented students. Every year more than 10 million Chinese students participate in a math competition, compared to the half-million or so American students who take an AMC exam. According to Zuming

Feng, who grew up in China before immigrating to the United States, "In China they have teachers in many high schools who are as devoted to competitions as Titu and I are. That is what we would need to be that strong."

Mathematical ability also figures prominently in China's college entrance exam, which features three to five mathematical problems structured as proofs. Chinese students therefore have a strong incentive to learn how to do Olympiad-style problems.

Finally, the Chinese Olympians' training is as rigorous as any in the world. Though they do not attend a summer training camp, the team members are selected through at least ten Olympiad-level tests designed to toughen them up for the competition.

"I follow sports," Titu once said when asked why the Chinese team is so good, "and there was this Ping-Pong player from China who won the world championship, but nobody had ever heard of him. The reporters came up to him afterward and asked, 'How was this world championship?' And he said, 'It was easy. The hard part was making the Chinese team.'"

▽

As the scores for the second day began to be posted, it quickly became clear that no team was going to be able to catch the Chinese. They scored straight 7s on problems four and five and five 7s on problem six — an incredible second-day performance. China had won its third straight Olympiad. The informal contest among countries was now for second and third place. And for the individual competitors the question remained: who would win gold, silver, and bronze medals?

Medals at the Olympiad are awarded strictly on the basis of scores. The top twelfth of the finishers get gold medals; the next sixth receive silvers; the next quarter, bronzes. Thus, about half of the Olympians return home with medals.

At the Forty-second Olympiad everyone who scored more than 10 points — out of a total of 42 possible — received a medal. That's how difficult the problems were. Some competitors returned home without scoring a single point, despite being the best high school mathematicians in their countries.

Four of the 473 competitors received perfect scores — they solved all six problems correctly. Two of those four were on the U.S. team. Reid Barton and Gabriel Carroll both knew that they had aced the exam. After the test each day, when asked how they had done, they answered, "I got them all." Throughout the event they exuded a quiet confidence that both intimidated and inspired their teammates. They were seniors competing in their fourth and third Olympiad, respectively. They knew they would do well.

But no one expected them to do this well. A perfect score in an Olympiad is the Mount Everest of mathematical competitions. And for Reid Barton, in particular, his straight 7s were the culmination of the most storied run in Olympiad history. He had received a gold medal after his freshman year, his sophomore year, his junior year, and now his senior year. No other competitor from any country in the forty-two years of the Olympiad had ever won four straight gold medals.

The two other perfect scores were achieved by Liang Xiao and Zhiqiang Zhang of China. Mihai Manea of Romania scored 41 points — faltering only on problem three — and Sergey Spiridonov of Russia scored 39. No one else achieved a score higher than 37. The highest-scoring girl — Greta Panova of Bulgaria, who finished higher than all but nine people in the competition — received a 36.

▽

Titu has never spoken with anyone about the judging, and he refuses to do so to this day. He would prefer that the following story remain private. But enough people were present in the judging rooms that the story can be reconstructed, and it was a pivotal moment in the Forty-second International Mathematical Olympiad.

The final problem to be scored was problem six, and the final U.S. solution to be scored was Tiankai's. "His solution was brilliant, but he didn't finish it," says Zvezda Stankova. Problem six was the number-theory problem involving four numbers labeled a, b, c, and d. The competitors were told that the numbers solved a particular equation, and their task was to prove that $ab + cd$ was not a prime number. The problem can be solved in various ways. Both Gabriel and Reid used imaginary numbers, which enabled the expression $ab + cd$ to be factored in such a way that it could not be prime. An algebraic approach used a proof by contradiction. An especially ingenious geometric proof began by setting the numbers a, b, c, and d equal to the lengths of a four-sided polygon. But all the solutions were very difficult. Of the 473 competitors at the Olympiad, only about twenty solved problem six.

Tiankai's approach was different from everyone else's. He expressed the four numbers a, b, c, and d in terms of four other numbers, which he called j, k, m, and n. He then set out to prove that these new numbers, if multiplied together, yielded a non-prime product, which in turn implied that $ab + cd$ could not be prime. His approach would have worked if he had had more time. But the air horn blew before he could complete his calculations.

Clearly, he did not deserve to score a 7, because he had not solved the problem. The question was whether he deserved a 4, a 5, or a 6. For the U.S. team the panel of coordinators who would judge problem six consisted of the Bulgarian coaches. Stankova, who had the distinct advantage of being able to argue with the coordinators in Bulgarian rather than in translated English, presented Tiankai's case. The tide swept back and forth, with near agreements on 4 points, then on 6, and finally on 5. It seemed the fairest score, though a strong argument could be made that the brilliance of Tiankai's unusual approach deserved a 6.

Stankova agreed with the 5, but she was only an assistant

coach. Titu would have to approve the score before it would be final. Titu, meanwhile, had been calculating the collective scores for the teams. He knew that if Tiankai was awarded a 5, the U.S. team would tie with Russia for second place, with 196 points. If Tiankai received a 6, the United States would beat Russia by one point. Furthermore, a 6 would guarantee a gold medal for Tiankai. And if Tiankai received a gold medal in his freshman year, he would be in a position to become only the second Olympian in history, after Reid Barton, to win four consecutive gold medals.

Stankova and the coordinators presented Titu with the agreed-on score of 5. Titu knew that if he insisted on having it revisited, he might get a 6. But the coordination was running late, the awards ceremony was the next day, and at that moment a tie with Russia seemed the best possible outcome. "Five's fine," Titu said. And with those words the scoring of the Forty-second Olympiad was over.

▽

The awards ceremony was held the next day in the opera hall of the Kennedy Center for the Performing Arts, which is on the northern shore of the Potomac River between the Watergate apartments and the Lincoln Memorial. The ceremony was funded largely by the Clay Mathematics Institute, and short versions of the Millennium Problems were projected on a huge screen behind the speakers. Flags from the eighty-three countries represented at the Olympiad ringed the stage. Before the ceremony the musical theme of the Olympic Games filled the hall.

As coaches, parents, and teammates cheered from the balconies, the gold, silver, and bronze medalists trooped down the aisles and onto the stage. They bent their heads, and Andrew Wiles placed the medals around their necks. He looked slight and retiring, as he often does in public. But when he spoke to the hundreds of people in the audience, his voice did not waver. "Let me congratulate you all," he began. "Some have arrived here by overcoming immense personal difficulties, others have arrived here overcoming only immense mathematical difficulties, but all of you have shown great talent and a real capacity for tremendous hard work."

In his talk Wiles described an ancient mathematical problem involving right triangles whose sides have rational lengths. Some such triangles have areas that are equal to whole numbers, whereas others don't. No one has yet been able to prove which whole numbers are associated with such triangles and which aren't. Whoever produces such a proof will be well on the way to solving the Millennium Problem called the Birch and Swinnerton-Dyer conjecture and earning \$1 million.

Then Wiles offered some advice to the Olympians. He encouraged them to consider careers in mathematics but cautioned that solving Olympiad problems is not like doing mathematical research and is not necessarily the best training for research. Working at the mathematical frontiers is more like a marathon than a sprint. Problems can take many years to solve, and you never know for sure whether you're going to reach the finish line. "The transition from a sprint to a marathon requires a new kind of stamina and a profoundly different test of character," he said. "We admire someone who can win a gold medal in four successive Olympic Games, not so much for the raw talent as for the strength of will and determination to pursue a goal over such a sustained period of time. Real mathematical theorems will require the same stamina whether you measure the effort in months or in years. You can forget the idea, if you ever had it, that all you require is a bit of natural genius and that then you can wait for inspiration to strike. There is simply no substitute for hard work and perseverance."

▼

So what about the question of genius, which is where this book began? Are the workings of the Olympians' minds incomprehensible, meaning that we must ascribe their insights to divine inspiration? Or is their problem solving a logical extension of the thinking people do every day?

One important observation is that the Olympians always had a reason for doing what they did, even when they let their minds wander. They never blundered about wildly until they happened across an approach that would work. They knew where they needed to go, and they moved forward based on past successes.

Sometimes their progress depended on their ability to combine separate skills. They might have blended deep intuition with rigorous logic, or they might have applied an old technique in a new context. But even when they combined two or more elements in a solution, those elements were not exotic or unknowable; the Olympians used tools familiar to everyone. In that respect their actions were not at all inexplicable.

Examining what they did, however, yields only part of the answer. At a deeper and more fundamental level, a profound mystery does remain. How could the Olympians see an image in their minds with such clarity that it became real? How did they know the best way to dissect a problem so that its constituent pieces suggested a solution? Why were they able to persevere with approaches that others would consider hopeless, until finally a solution emerged?

These mysteries are not confined to the solving of difficult mathematical problems. They are familiar mysteries — the kinds of questions people ask all the time. How do we construct order out of chaos? How do our minds create something that has never existed before? Why do some things attract our interest while others leave us cold? The achievements of the Olympians do not necessarily derive from obscure mental abstractions. They revolve around the much more immediate and pragmatic considerations of how humans make sense of the world.

In that respect, the genius we sometimes sense in the Olympians is not something mysterious and unknowable; it is present in everyone in some form. When we perceive the order in an abstract painting, we do not ask what makes that perception possible. When we pursue a task with dedication and purpose, we do not stop to question our motivations. When we walk down a wooded path and sense, just for a moment, how amazing it is to live in this world of trees and sky and rich brown earth, we are experiencing the same emotions mathematicians feel when they discover something they never knew was there.

Maybe that's why the idea of genius appeals to us so strongly — because it partakes of a mystery with which we all are familiar. The sense of mystery may resonate in us most deeply when we are admiring creative works of great beauty and power. But the emotions evoked by these great achievements are not foreign to any of us. On the contrary, we feel those emotions because great achievements spring from abilities that we all, in our own way, share.

▽

After the awards ceremony was over, the Olympians spilled from the Kennedy Center's concourse onto a terrace overlooking the Potomac River. It was a pleasant Friday afternoon, and the water was thronged with pleasure boats. An eight-man shell passed silently along the river like a water strider on a pond. Several military helicopters flew upriver, carrying the president to Camp David for the weekend. A few minutes later a steady stream of jets resumed their downriver approaches to National Airport.

The Olympians and their coaches milled about in groups, having their pictures taken, showing each other their medals. Some asked for an autograph from Andrew Wiles, who stood on the edge of the gathering. Others were saying goodbye to their guides, since this was the moment when, after ten straight days, the guides and their teams had to separate.

The next day the Olympians would board planes to go

home. They would return to noisy celebrations and to indiffer-
ence, to happy families and to turmoil, to repressive regimes and
to peaceful democracies, to fulfilling careers and to disappoint-
ments. But for this moment they were united in their achievement. They had competed in the Forty-second International
Mathematical Olympiad. No one could ever take that away from
them.

11 · epilogue

After the Olympiad, Melanie Wood returned for her third year at Duke University in Durham, North Carolina. She had already decided to major in math, and her upper-level classes consumed much of her time. But she stayed in touch with a wide circle of friends and spoke whenever she could to groups of girls interested in math. Following her senior year she became a Gates Scholar and traveled to Cambridge, England, to study math for a year, after which she planned to attend graduate school at Princeton.

In 2003 coach Titu Andreescu decided that it was time to do something new. He applied for several academic positions and took a job at the University of Wisconsin at Whitewater while serving on the 2003 Olympiad advisory board and on the problem selection and judging committees. Zuming Feng, the assistant coach at the Forty-second Olympiad, was team leader at the 2003 Olympiad in Tokyo.

Tiankai Liu earned a gold medal at the Olympiad in Glasgow in 2002. But the summer after his junior year he decided to skip the Olympiad in Tokyo. Instead he attended the Research Summer Institute at MIT, which is a common steppingstone for rising high school seniors who hope to be finalists in the Intel Science Talent Search, the prestigious national science fair that has launched the careers of many prominent scientists. For now, Reid Barton's record of four straight gold medals is safe.

Ian Le entered Harvard College the fall after the Olympiad. A math major, he remained active in music and began taking private piano lessons again.

David Shin entered MIT, where he began working on a double major in mathematics and computer science. In his junior year, having continued to play jazz piano throughout college, he decided to minor in music as well.

Oaz Nir joined Melanie at Duke. Like David, he declared a double major, but in mathematics and English. "I think I'll go to graduate school," he said, "but I don't know in what."

Reid Barton entered MIT the fall after the Olympiad. That December he participated in the William Lowell Putnam competition, the premier mathematics competition for college students in the United States and Canada. He was one of the top five finishers (individual positions among the top five are not announced), and the team from MIT finished second only to Harvard.

Gabriel Carroll was on Harvard's Putnam team as a college freshman and was also among the top five individual scorers on the test. Over the course of his first year at Harvard, his Web site continued to grow. "Was it a significant year?" he wrote in the spring. "Time alone will tell. . . . I could not possibly have predicted four years ago, or two years ago, or even one year or six months ago what my life would look like now; changes are incessant and to a large extent random. So I simply acknowledge this and wait to see."

appendix
Solutions and Commentaries

ODDS, EVENS, AND SQUARES (CHAPTER 1)

The problem on page 35 — how many of the integers between 1 and 1,000, inclusive, can be expressed as the difference of the squares of two nonnegative integers — appeared on the 1997 American Invitational Mathematics Examination (AIME).

In mathematical terms, if x is a whole number between 1 and 1,000 (inclusive), can we find two other whole numbers — let's call them a and b — such that $x = a^2 - b^2$?

The odd numbers are easier to account for than the even numbers. If x is an odd number, it can be written as $2n + 1$ for some whole number n. (Thus, if x is 7, n is 3.) Now consider the number $(n + 1)^2$, which is equal to $(n + 1)(n + 1)$, or $n^2 + 2n + 1$. We can use that equation to express $2n + 1$ in a different way. We can write $2n + 1 = n^2 + 2n + 1 - n^2 = (n + 1)^2 - n^2$. But $2n + 1$ is equal to x, so if we let $n + 1 = a$ and $n = b$, then $x = a^2 - b^2$. This proves that all the odd numbers from 1 to 1,000 (in fact, all odd numbers to infinity) can be expressed as the difference between two squares. Most problem solvers recognize this right away because they know that the square numbers $(1, 4, 9, 16 \ldots)$ are separated by successive odd numbers.

The even numbers are a bit tougher. First, any good high

school math competitor knows that $a^2 - b^2 = (a + b)(a - b)$. Now think what happens when a and b are both even, both odd, or one of each. If they are both even, then $(a + b)$ and $(a - b)$ are both even, and $(a + b)(a - b)$ is also even. If they are both odd, then $(a + b)$ and $(a - b)$ are still both even, as is their product. If one number is odd and the other is even, then both $(a + b)$ and $(a - b)$ are odd, and so is $(a + b)(a - b)$. We can ignore this option, because we've already taken care of the odd numbers.

Any product of two even numbers has to be divisible by 4 without leaving a remainder. (Some experimenting will demonstrate this fact. Technically, it occurs because the product will have at least two 2s in its prime factorization.) But the even numbers that cannot be evenly divided by 4 — 2, 6, 10, and so on — cannot be expressed as the product of two even numbers. These numbers therefore cannot be expressed in the form $(a + b)$ $(a - b)$, which also means that they cannot be expressed as the difference of two squares.

But is the converse — that all the even numbers divisible by 4 (that is, 4, 8, 12, and so on) *can* be expressed as the difference of two squares — necessarily true? Yes, because if x is a multiple of 4, it equals $4n$ for some number n. But $4n$ can also be written as $(n + 1)^2 - (n - 1)^2$. So if $a = (n + 1)$, $b = (n - 1)$, and $x = 4n$, then $x = a^2 - b^2$.

What this proof shows is that half of all the even numbers (those divisible by 4) can be expressed as the difference between two squares. The first 1,000 whole numbers contain 500 even numbers. So half of those numbers, plus all the odd numbers, can be expressed as the difference between two squares, and the answer to the problem is 750.

By the way, this problem was the first and easiest on the 1997 AIME. The other fourteen problems on the three-hour test were appreciably harder.

A Mathematical Chestnut (Chapter 2)

Explaining the humor of a mathematical joke is as graceless as explaining any other joke, but here goes: People who took calculus in high school or college often forget that when an expression like x^2 is integrated, the answer technically includes a constant of integration. The waitress in Gabriel's story knew calculus so well that she improved on the answer she was given by the mathematician.

Problem One — A Proof by Contradiction (Chapter 3)

Tiankai's proof by contradiction in his solution to problem one was a marvel of mathematical concision. He had already shown that angle PAO in the following diagram is equal to or greater than 30 degrees. Now he had to use that information to figure out where point P is located between points C and M. If he could prove that point P is closer to C than it is to M, it would follow that angle CAB plus angle COP is less than 90 degrees, which is what he had to prove.

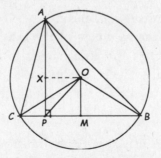

First he assumed the opposite. He said, "Let's assume that P is closer to M than it is to C." But PM is the same distance as XO in the diagram Tiankai drew, because XOMP is a rectangle. Also, because Tiankai had demonstrated that angle PAO was at least 30 degrees, he knew that the sides of triangle XAO are related in a particular way. In a 30-60-90–degree triangle, the shortest side is always half the length of the longest side. Therefore, XO was at least half the length of AO. And because AO is the radius of the circle, XO had to be at least half the radius of the circle.

Now look at the distance CM. That distance has to be less than the radius of the circle, because CM is always less than CO. So if XO, which is equal to PM, is more than half the radius of the circle, and if CM is less than the radius of the circle, P has to be closer to C than to M. The assumption Tiankai had made had to be incorrect.

There are other ways to prove this point, but the Olympians often used proofs by contradiction in their solutions to the problems.

Problem Two — Jensen's Inequality (Chapter 4)

To solve problem two, Ian had to prove that

$$a/\sqrt{a^2 + 8bc} + b/\sqrt{b^2 + 8ac} + c/\sqrt{c^2 + 8ab} \geq 1$$

for any positive numbers a, b, and c.

Alone among the members of the U.S. team, Ian realized that he could prove this using the mathematical equation known as Jensen's inequality. The inequality looks daunting but is not really that complicated. Here it is for three variables (in this equation, $r + s + t$ has to equal 1):

$$r f(x) + s f(y) + t f(z) \geq f(rx + sy + tz)$$

The letters x, y, and z and r, s, and t stand for numbers. But the letter f, which is short for function, represents a mathematical action in which one does something to the number represented by x, y, or z (such as doubling or squaring it) to get another number. Jensen showed that the above inequality holds for functions that curve upward when plotted on graph paper.

In his solution to problem two, Ian let the function, f, in Jensen's inequality be

$$f(x) = 1/\sqrt{x}$$

The graph of this function curves upward, so Ian knew that the inequality would hold. He then had to define x, y, and z in the inequality. That was fairly easy, given that he had already established that f would stand for taking the square root of a number and dividing 1 by the result. He set the x of Jensen's inequality to $a^2 + 8bc$; y to $b^2 + 8ac$; and z to $c^2 + 8ab$. That way, $f(x)$ was equal to

$$1/\sqrt{a^2 + 8bc}$$

and so on for y and z. So

$$f(x) + f(y) + f(z) = 1/\sqrt{a^2 + 8bc} + 1/\sqrt{b^2 + 8ac} + 1/\sqrt{c^2 + 8ab}$$

Now Ian did something extremely clever. He set the numbers r, s, and t in Jensen's inequality to be $a/(a + b + c)$, $b/(a + b + c)$, and $c/(a + b + c)$. If you put all those numbers into the inequality and do some algebra, you come up with the following equation, which Ian jotted down on his scratch paper:

$$\frac{\frac{a}{a+b+c}}{\sqrt{a^2 + 8bc}} + \frac{\frac{b}{a+b+c}}{\sqrt{b^2 + 8ac}} + \frac{\frac{c}{a+b+c}}{\sqrt{c^2 + 8ab}} \geq \frac{\sqrt{a+b+c}}{\sqrt{a^3 + b^3 + c^3 + 24abc}}$$

The term $a + b + c$ appears on both sides of the equation, so you can multiply both sides of the equation by that term to get

$$\frac{a}{\sqrt{a^2+8bc}}+\frac{b}{\sqrt{b^2+8ac}}+\frac{c}{\sqrt{c^2+8ab}}\geq\frac{\sqrt{(a+b+c)^3}}{\sqrt{a^3+b^3+c^3+24abc}}$$

Now the left-hand side of the equation is the same as the quantity specified in problem two. As described in Chapter 4, Ian easily demonstrated that the right-hand side of the equation had to be more than 1. Therefore the left-hand side of the equation also had to be more than 1, and his proof was complete.

PROBLEM THREE — THE AXIOM OF CHOICE (CHAPTER 5)

Only one other writer attended the Forty-second Olympiad — a California freelancer named Dana Mackenzie, who has a Ph.D. in mathematics and was covering the event for *Science* magazine. A few days after the end of the Olympiad, he sent me the following e-mail message:

> I have one little story to tell you about [problem three]. The mathematical humor involved is a bit too subtle for my short article in *Science,* but it might give your readers some insight into what math students laugh about.
>
> As you already know, the first step in Reid's proof was to make a 21 × 21 table, listing the girls down the side and the boys across the top. For the entry in the ith row and jth column of the table, you choose "one" problem that was solved by both girl i and boy j. When Reid explained his solution to me the first time, I didn't catch this subtle point, and I thought that he said to list "all" of the problems that were solved by both girl i and boy j. It was only a couple of days later, when I was writing the solution up for *Science,* that I realized that the last step of the proof doesn't work if you do it that way.
>
> Now I was panicking, because I had promised my editor

to send him the solution to this problem! That's when I was saved by something that Gabriel [another team member] had said. After Reid had explained his solution to me, Gabe had razzed Reid for "using the Axiom of Choice 441 times." The Axiom of Choice, as you might know, is the statement that if you have any collection of sets (say, baskets of eggs), then you can talk about a set created by arbitrarily selecting one element from each set (that is, one egg from each basket). It sounds obvious, but historically there has been a lot of debate in the mathematical community over whether you can apply this axiom when you have an infinite number of sets. Some very strange paradoxes arise if you allow this. Most mathematicians do allow it, but a minority of mathematicians consider it to be an invalid procedure. Partly for this reason, even mathematicians who accept the "infinite" form of the Axiom of Choice view such proofs as slightly tainted. It's poor form to use it if you don't need to.

Anyway, I did not understand Gabe's comment at the time, but two days later its meaning hit me. Gabe was referring to the fact that Reid's proof depended on arbitrarily choosing one problem to write in each of the 441 cells in the table from the possibly several questions that the corresponding pair of students got right. And that was precisely what he needed to make the last step work. Now it all made sense!

This episode was very impressive to me. First, I was amazed by Reid's insight in avoiding a trap that probably would have caught me, even if I had been clever enough to come up with the rest of the proof. Second, even as high school students, they've heard about the Axiom of Choice, and moreover they've absorbed the lore enough to know that mathematicians view it with some disfavor. That's why Gabe was razzing Reid. It's like saying, "Yeah, you won the race, but your shoelaces were untied." I'm sure Gabe knew that using the Axiom of Choice wasn't a "real" objection to Reid's proof, but he was teasing him about it all the same because it made Reid's proof seem a little less elegant. In reality, of course, Reid's proof is very elegant.

Problem Four — The Sum of All Sums (Chapter 7)

As with many of the other proofs on the Olympiad, David's proof for problem four proceeds by contradiction. The problem defines the sum

$$S(a) = \sum_{i=1}^{n} c_i a_i$$

in which the term a represents a distinct ordering of the first n whole numbers. There are n factorial, or $n!$, such permutations, so a can take $n!$ forms. The n numbers represented by c_i, in which i ranges from 1 to n, are simply integers. So $S(a)$ represents a number, though it could be a very large one.

To solve the problem, you have to prove that there exist distinct permutations of the first n whole numbers (the problem calls these permutations b and c) such that $n!$ evenly divides $S(b) - S(c)$ when n is an odd number. To produce a contradiction, you first assume that the statement you want to prove is false. So you assume that no permutations b and c exist such that $n!$ evenly divides $S(b) - S(c)$. You then add up all the $S(a)$'s over the $n!$ permutations of a, which yields

$$\sum_{a} \sum_{i=1}^{n} c_i a_i$$

Through some fancy calculating, you can show that this sum cannot be evenly divided by $n!$ if the assumption you made is correct. But you can also calculate the sum another way, and when you do that, you discover that the sum *can* be evenly divided by $n!$ when n is an odd number. Therefore the assumption that $S(b) - S(c)$ could not be divided by $n!$ for some b and c when n is odd is incorrect, and the proof is done.

PROBLEM FIVE — THE MYSTERIOUS CONSTRUCTION (CHAPTER 8)

The following riff on Oaz's solution, which was developed by one of the Chinese Olympians, illustrates an extremely clever way to solve problem five, in which you have to calculate the angles of triangle ABC. You already know that angle A is 60 degrees and that line AP bisects the angle. You also know that line BQ bisects angle B, with each half of the angle denoted by beta (β). And you know that the lengths of various line segments meet the following condition: AB + BP = AQ + QB. And that's all you know.

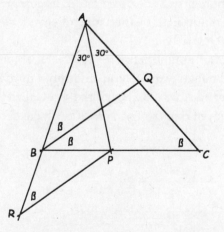

In this solution, you extend line AB to the point called R, and you make BR equal in length to BP. Using the condition stated in the problem that AB + BP = AQ + QB, you can prove (with some difficulty) that AR = AC. Now look at triangle BRP.

It's an isosceles triangle, with the big angle equal to 180 degrees − 2β. Therefore angle BRP has to equal beta. But if angle BRP equals beta, then so does angle ACB, because line AP bisects angle A, which means that the two halves of the chevron formed by points A, R, P, and C are identical. If angle C equals beta, and angle B equals two times beta, then beta has to equal 40 degrees for the angles of triangle ABC to add to 180 degrees.

It's a beautiful proof, but how would anyone know to extend line AB to R, especially under the time pressures of an Olympiad? That's the mystery.

Problem Six — An Imaginary Factor (Chapter 9)

Gabriel's use of imaginary numbers in problem six was directly linked to the famous equation $e^{\pi i} = -1$. The number omega (ω) is defined as $\omega = e^{2\pi i/3}$. So $\omega^2 = e^{2\pi i/3} \times e^{2\pi i/3} = e^{2\pi i/3 + 2\pi i/3} = e^{4\pi i/3}$ (because the exponents of e can be added together when the two numbers are multiplied). By the same token, $\omega^3 = e^{6\pi i/3} = e^{2\pi i} = e^{\pi i} \times e^{\pi i} = -1 \times -1 = 1$.

Thus the set of numbers 1, −1, ω, −ω, ω^2, and −ω^2 are related in a particular way. If you multiply any two of them together, you get another member of the set. Gabriel used the powerful properties of this group to crack problem six.

sources

Unless otherwise noted, all the quotations in this book are from interviews. Listed thematically below are the written sources used for each chapter.

INTRODUCTION
The article by Nura Turner that helped convince U.S. mathematicians to send a team to the International Mathematical Olympiad is "Why Can't We Have a USA Mathematical Olympiad?" *American Mathematical Monthly* 78 (1971): 192–95. A later article by Turner described the early years of U.S. participation: "A Historical Sketch of Olympiads: U.S.A. and International," *College Mathematics Journal* 16 (1985): 330–35.

Descriptions of the problems and summaries of the results for the first twenty-seven Olympiads can be found in *International Mathematical Olympiads 1959–1977* by Samuel Greitzer (Washington, D.C.: Mathematical Association of America, 1978) and in *International Mathematical Olympiads and Forty Supplementary Problems 1978–1985* by Murray Klamkin (Washington, D.C.: Mathematical Association of America, 1986).

Marjorie Garber's description of geniuses appears in "Our Genius Problem," *Atlantic Monthly* (December 2002), pp. 65–72. Mark Kac divides geniuses into ordinary and magical in *Enigmas of Chance: An Autobiography* (New York: Harper and Row, 1985).

Alex Kasman's reviews of fiction involving mathematics appear at   http://math.cofc.edu/faculty/kasman/MATHFICT/default.html.

The "Math in the Movies" Web page, at http://world.std.com/
~reinhold/mathmovies.html, offers a similar but smaller list. A
DVD of *Fermat's Last Tango* is available from the Clay Mathematics Institute, One Bow St., Cambridge, MA 02138.

Solly Zuckerman's treatment of mathematical equations is recounted in the first chapter of *The Doctrine of DNA: Biology as Ideology*, by Richard Lewontin (New York: Penguin, 1993).

The World of Mathematics, compiled by James Newman (New
York: Simon and Schuster, 1956; reprint, Dover, 2000), is a four-
volume "library of the literature of mathematics."

Chapter 1. Inspiration
The best source of information about Mathcounts is the program's
Web site: http://mathcounts.org. The organization's office is at 1420
King St., Alexandria, VA 22314.

Melanie Wood is profiled by Polly Shulman in "The Girl Who
Loved Math," *Discover* (June 2000), pp. 67–70. Melanie reflects
on her Mathcounts experiences in *MATHCOUNTS News* 17 (Fall
2001), p. 9. The book she read after her initial success at Mathcounts is *The Art of Problem Solving*, by Sandor Lehoczky and
Richard Rusczyk (New York: Greater Testing Concepts, 1993).

Camilla Persson Benbow and Julian Stanley's "Sex Differences in Mathematical Ability: Fact or Artifact?" was in *Science* 210
(1980): 1262–64. Letters in response appear in *Science* 212 (1981):
114–21. For an analysis of the controversy that arose over that
article, see "Social Forces Shape Math Attitudes and Performance," by Jacquelynne Eccles and Janis Jacobs, *Signs* 11 (1986):
367–80. A more policy-oriented article by Benbow and Stanley is
"Inequity in Equity: How 'Equity' Can Lead to Inequity for High-
Potential Students," *Psychology, Public Policy, and Law* 2 (1996):
249–92.

Several papers from a 1992 conference honoring Stanley appear in *Intellectual Talent: Psychometric and Social Issues,* edited
by Camilla Persson Benbow and David Lubinski (Baltimore: Johns
Hopkins University Press, 1996). Information about the Center for
Talented Youth at Johns Hopkins can be found on the Web at http://

www.cty.jhu.edu/cde. A good overview of the longitudinal study codirected by Benbow is "The Study of Mathematically Precocious Youth: The First Three Decades of a Planned 50-Year Study of Intellectual Talent," by David Lubinski and Camilla Persson Benbow, pp. 251–81 in *Beyond Terman: Contemporary Longitudinal Studies of Giftedness and Talent,* edited by Rena Subotnik and Karen Arnold (Norwood, N.J.: Ablex, 1994).

Benbow, Lubinski, and several coauthors report on the career choices of the individuals tracked through SMPY in "Sex Differences in Mathematical Reasoning Ability at Age 13: Their Status 20 Years Later," *Psychological Science* 11 (2000): 474–80; and in "Top 1 in 10,000: A 10-Year Follow-Up of the Profoundly Gifted," *Journal of Applied Psychology* 86 (2001): 718–29.

The most detailed and specific information about gender differences in math ability can be found in "Sex Differences in Mathematical Reasoning Ability in Intellectually Talented Preadolescents: Their Nature, Effects, and Possible Causes," by Camilla Benbow, *Behavioral and Brain Sciences* 11 (1988): 169–232; and "Sexual Selection and Sex Differences in Mathematical Abilities," by David Geary, *Behavioral and Brain Sciences* 19 (1996): 229–84. The articles are accompanied by responses from a variety of perspectives.

A critique of the education girls receive is *How Schools Short-change Girls — the AAUW Report* (New York: Marlowe, 1995). "Environmental Input to the Development of Sex-Related Differences in Spatial and Mathematical Ability," by Maryann Baenninger and Nora Newcombe, *Learning and Individual Differences* 7 (1995): 363–79, documents the different experiences of boys versus girls in activities that might affect mathematical ability.

Many books and articles look more broadly at possible cognitive differences between males and females, including *Myths of Gender: Biological Theories About Women and Men,* 2nd ed., by Anne Fausto-Sterling (New York: Basic Books, 1985); *Sex and Cognition,* by Doreen Kimura (Cambridge, Mass.: MIT Press, 2000); and *Sex Differences in Cognitive Abilities,* 3rd ed., by Diane Halpern (Hillsdale, N.J.: Erlbaum, 2000).

Women and the Mathematical Mystique, edited by Lynn Fox,

Linda Brody, and Dianne Tobin (Baltimore: Johns Hopkins University Press, 1980), analyzes many of the social pressures that discourage women from pursuing mathematical interests. *Women in Mathematics: The Addition of Difference,* by Claudia Henrion (Bloomington: Indiana University Press, 1997), profiles professional women mathematicians and examines why their representation in the field remains so low.

The AMC 8, 10, and 12 tests are administered by the Mathematical Association of America's American Mathematics Competitions, University of Nebraska, P.O. Box 81606, Lincoln, NE 68501. The AMC Web site is http://www.unl.edu/amc.

Paul Zeitz presents problems ranging from the straightforward to the exceedingly difficult in *The Art and Craft of Problem Solving* (New York: Wiley, 1999). Many books of problems from national Olympiads and other competitions have been published—for example, *Mathematical Olympiads: Problems and Solutions from Around the World, 1999–2000,* by Titu Andreescu and Zuming Feng (Washington, D.C.: Mathematical Association of America, 2001). Some of Titu's favorite problems are included in *Mathematical Miniatures,* by Svetoslav Savchev and Titu Andreescu (Washington, D.C.: Mathematical Association of America, 2002).

At the opposite end of the spectrum, two excellent introductions to mathematics and problem solving for elementary school children are *The Number Devil,* by Han Magnus Enzensberger (New York: Metropolitan, 1998); and *Sideways Arithmetic from Wayside School,* by Louis Sachar (New York: Scholastic, 1992).

CHAPTER 2. DIRECTION

Michael Mahoney describes Fermat's life and mathematics in *The Mathematical Career of Pierre de Fermat,* 2nd ed. (Princeton, N.J.: Princeton University Press, 1994).

For a brief history of *KöMal* and problem solving in eastern Europe, see *C2K: Century 2 of Kömal,* by George Berzsenyi (Budapest: Roland Eötvös Physical Society, 1999).

The videotape "Eighth-Grade Mathematics Lessons: United

States, Japan, and Germany" is available from the superintendent of documents, P.O. Box 371954, Pittsburgh, PA 15250–7954 (or call 202-512-1800 in Washington, D.C.). For more information on this study, see "Candid Camera" by Steve Olson, *Teacher Magazine* (June 1999), pp. 28–32.

James Stigler and James Heibert present their ideas about how to improve the quality of mathematics education in *The Teaching Gap: Best Ideas from the World's Teachers for Improving Education in the Classroom* (New York: Free Press, 1999). Stigler, Heibert, and Ronald Gallimore elaborate on these ideas in "A Knowledge Base for the Teaching Profession: What Would It Look Like and How Can We Get One?" *Educational Researcher* 31, no. 5 (2002): 3–15.

The National Council of Teachers of Mathematics offers a comprehensive and balanced approach to the teaching of mathematics in *Principles and Standards for School Mathematics* (Reston, Va.: National Council for Teachers of Mathematics, 2000). A short booklet that resolves the major issues at stake in the "Math Wars" is *Helping Children Learn Mathematics,* edited by Jeremy Kilpatrick and Jane Swafford (Washington, D.C.: National Academy Press, 2002).

CHAPTER 3. INSIGHT

A good one-volume history of mathematics is *The Rainbow of Mathematics: A History of the Mathematical Sciences,* by Ivor Grattan-Guinness (New York: W. W. Norton, 1997).

The article Tiankai's parents read was "Math Mania," by Glennda Chui, *San Jose Mercury News* (October 27, 1998), p. F1.

The youthful exploits of Carl Friedrich Gauss were recounted the year after his death by his colleague Wolfgang Sartorius von Walterschausen in *Gauss: A Memorial* (Leipzig: S. Hirzel, 1856).

Sylvia Nasar discusses John Nash's preoccupation with age in chapter 32 of *A Beautiful Mind* (New York: Simon & Schuster, 1998). The quotation by G. H. Hardy on age is from *A Mathematician's Apology* (1940; reprint, Cambridge University Press, 1992).

Nancy Stern analyzes the productivity of mathematicians at different ages in "Age and Achievement in Mathematics: A Case-Study in the Sociology of Science," *Social Studies of Science* 8 (1978): 127–40.

The paper that resulted from Roger Shepard's 1968 vision is "Mental Rotation of Three-Dimensional Objects," *Science* 171 (1971): 701–3. Shepard reviews the role played by mental imagery in science, mathematics, and engineering in "Externalization of Mental Images and the Act of Creation," pp. 133–89 in *Visual Learning, Thinking, and Communication,* edited by Bikkar Randhawa and William Coffman (New York: Academic Press, 1978). A compilation of papers documenting the development of these ideas appears in *Mental Images and Their Transformation,* edited by Roger Shepard and Lynn Cooper (Cambridge, Mass.: MIT Press, 1982). Shepard writes in a more popular vein in *Mind Sights: Original Visual Illusions, Ambiguities, and Other Anomalies, with a Commentary on the Play of Mind in Perception and Art* (New York: W. H. Freeman, 1980). For an extension of Shepard's work to other perceptual systems, see "Perceptual-Cognitive Universals as Reflections of the World," by Roger Shepard and respondents, *Behavioral and Brain Sciences* 24 (2001): 581–601.

The illustration is from an adaptation of Shepard's work by Steven Vandenberg and Allan Kuse, "Mental Rotations, A Group Test of Three-Dimensional Spatial Visualization," *Perceptual and Motor Skills* 47 (1978): 599–604.

Michael Corballis provides a brief history of research on mental rotation in "Mental Rotation and the Right Hemisphere," *Brain and Language* 57 (1997): 100–121. An overview of the role of mental imagery in thought can be found in *Image and Brain: The Resolution of the Imagery Debate,* by Stephen Kosslyn (Cambridge, Mass.: MIT Press, 1994). A classic analysis of the role of mental imagery in mathematical thinking appears in *The Psychology of Mathematical Abilities in Schoolchildren,* by Vadim Andreevich Krutetskii (Chicago: University of Chicago Press, 1976).

Ellen Winner and Beth Casey describe their work on "visualizers" and "verbalizers" in "Cognitive Profiles of Artists," pp. 154–

69 in *Emerging Visions of the Aesthetic Process: Psychology, Semiology, and Philosophy*, edited by Gerald Cupchik and János László (New York: Cambridge University Press, 1992).

Beth Casey, Ronald Nuttall, and Elizabeth Pezaris explain the rationale behind their research on spatial reasoning in "Evidence in Support of a Model That Predicts How Biological and Environmental Factors Interact to Influence Spatial Skills," *Developmental Psychology* 35 (1999): 1237–47. For Casey's views on gender differences in spatial imagery, see "Understanding Individual Differences in Spatial Ability Within Females: A Nature/Nurture Framework," *Developmental Review* 16 (1996): 241–60.

Two popular books that discuss Paul Erdős's views about "the Book" are *My Brain Is Open: The Mathematical Journeys of Paul Erdős*, by Bruce Schechter (New York: Simon & Schuster, 1998); and *The Man Who Loved Only Numbers: The Story of Paul Erdős and the Search for Mathematical Truth*, by Paul Hoffman (New York: Hyperion, 1998).

CHAPTER 4. COMPETITIVENESS
Several more or less consistent histories of Ultimate Frisbee can be found on the World Wide Web. See, for example, http://www .northbayultimate.org/html/about/history.shtml.

For an interesting discussion of the role of perseverance in great achievement, see "The Genetics of Genius," by David Lykken, pp. 15–38 in *Genius and the Mind: Studies of Creativity and Temperament in the Historical Record*, edited by Andrew Steptoe (New York: Oxford University Press, 1998).

General reviews of psychological theories of motivation can be found in *Development of Achievement Motivation*, edited by Allan Wigfield and Jacquelynne Eccles (San Diego: Academic Press, 2002); and *Motivation: A Biosocial and Cognitive Integration of Motivation and Emotion*, by Eva Dreikurs Ferguson (New York: Oxford University Press, 2000). A shorter review is "Motivational Beliefs, Values, and Goals," by Jacquelynne Eccles and Allan Wigfield, *Annual Review of Psychology* 53 (2002): 109–32.

Alfie Kohn argues against competition in *No Contest: The Case Against Competition,* rev. ed. (Boston: Houghton Mifflin, 1992). Jock Abra disputes Kohn's conclusions in "Competition: Creativity's Vilified Motive," *Genetic, Social, and General Psychology Monographs* 119 (1993): 289–342.

Pamela Clinkenbeard analyzes some of the negative effects of competition, even on winners, in "The Motivation to Win: Negative Aspects of Success at Competition," *Journal for the Education of the Gifted* 12 (1989): 293–305. "Academic Competitions in Science: What Are the Rewards for Students?" by Tammy Abernathy and Richard Vineyard, *The Clearing House* 74 (2001): 269–76, is also pessimistic about the effects of competitions.

Two articles that examine the value of competition are "Why Do People Like Competition: The Motivation for Winning, Putting Forth Effort, Improving One's Performance, Performing Well, Being Instrumental, and Expressing Forceful/Aggressive Behavior," by Robert Franken and Douglas Brown, *Personality and Individual Differences* 19 (1995): 175–84; and "Competition," by John Wilson, *Journal of Moral Education* 18 (1989): 26–31.

James Campbell, Harald Wagner, and Herbert Walberg describe academic competitions for advanced students in "Academic Competitions and Programs Designed to Challenge the Exceptionally Talented," in *International Handbook of Giftedness and Talent,* edited by Kurt Heller, Franz Mönks, Robert Sternberg, and Rena Subotnik (New York: Elsevier, 2000). Competition specifically for talented students is examined in "Competition and the Adjustment of Gifted Children: A Matter of Motivation," by Stephen Udvari, *Roeper Review* 22 (2000): 212–17. For an analysis of educational preferences among good students, see "Gifted Secondary Students' Preferred Learning Style: Cooperative, Competitive, or Individualistic?" by Anita Li and Georgina Adamson, *Journal for the Education of the Gifted* 16 (1992): 46–54.

The information on Johan Ludvig William Valdemar Jensen is from the *Dictionary of Scientific Biography* (New York: Scribner's, 1973).

CHAPTER 5. TALENT

Ellen Winner's research is presented most thoroughly in *Gifted Children: Myths and Realities* (New York: Basic Books, 1996). See also her articles "The Origins and Ends of Giftedness," *American Psychologist* 55 (2000): 159–69; "Exceptionally High Intelligence and Schooling," *American Psychologist* 52 (1997): 1070–81; and "Giftedness and Egalitarianism in Education: A Zero Sum?" *NASSP Bulletin* 82 (1998): 47–60.

Other books on unusual early talent include *Nature's Gambit: Child Prodigies and the Development of Human Potential,* by David Feldman and Lynn Goldsmith (New York: Teachers College Press, 1991); *Exceptionally Gifted Children,* by Miraca Gross (London: Routledge, 1993); and *Child Prodigies and Exceptional Early Experience,* by John Radford (London: Harvester, 1990).

For Michael Howe's refutation of the "talent account," see *Genius Explained* (New York: Cambridge University Press, 1999) and the article by Howe, Jane Davidson, and John Sloboda, "Innate Talents: Reality or Myth?" *Behavioral and Brain Sciences* 21 (1998): 399–442.

K. Anders Ericsson and Neil Charness take positions similar to Howe's in "Expert Performance: Its Structure and Acquisition," *American Pscyhologist* 49 (1994): 725–47. Ericsson also writes about these ideas with Ralf Krampe and S. Heizman in "Can We Create Gifted People?" pp. 222–49 of *The Origins and Development of High Ability,* edited by Gregory Bock and Kate Ackrill (New York: Wiley, 1993); and with Ralf Krampe and Clemens Tesch-Römer in "The Role of Deliberate Practice in the Acquisition of Expert Performance," *Psychological Review* 100 (1993): 363–406.

The Origins and Development of High Ability contains a number of other interesting articles, including "The Concept of 'Giftedness': A Pentagonal Implicit Theory," by Robert Sternberg, pp. 5–21; "Psychological Profiles of the Mathematically Talented: Some Sex Differences and Evidence Supporting Their Biological Basis," by Camilla Persson Benbow and David Lubinski, pp. 44–66; "Genetics and High Cognitive Ability," by Robert Plomin and Lee Ann Thompson, pp. 67–84; and "The Early Lives of Child Prodigies," by Michael Howe, pp. 85–105. Another thought-provoking compilation is *Conceptions of Giftedness,* edited by Robert Sternberg and Janet Davidson (New York: Cambridge University Press, 1986).

Among the many biographies of Mozart are *Mozart: A Life,* by Maynard Solomon (New York: HarperCollins, 1995) and *Mozart,* by Peter Gay (New York: Viking, 1999).

A book that emphasizes the genetic contributions to talent is *Biological Approaches to the Study of Human Intelligence,* edited by Philip Vernon (Norwood, New Jersey: Ablex, 1993).

"The Development of Exceptional Research Mathematicians," by William Gustin, appears in *Developing Talent in Young People,* edited by Benjamin Bloom (New York: Random House, 1985). Clive Kilmister's "Genius in Mathematics" is in *Genius: The History of an Idea,* edited by Penelope Murray (New York: Basil Blackwell, 1989). Also see "Nurturing Talents/Gifts in Mathematics," by Wilhelm Wieczerkowski, Arthur Cropley, and Tania Prado, pp. 413–26 of *International Handbook of Giftedness and Talent,* 2nd ed., edited by Kurt Heller, Franz Mönks, Robert Sternberg, and Rena Subotnik (New York: Elsevier, 2000).

A biography of Blind Tom is *Blind Tom — The Black Pianist-Composer (1849–1908),* by Geneva Handy Southall (Lanham, Md.: Scarecrow Press, 1999). A good Web site about Blind Tom that includes the quotation from Mark Twain is at http://www.twainquotes.com/archangels.html. For a popular treatment of savants, see "Islands of Genius," by Darold Treffert and Gregory Wallace, *Scientific American* (June 2002), pp. 76–85.

David Moore offers a way to move beyond unproductive discussions of the origins of talent in *The Dependent Gene: The Fallacy of "Nature vs. Nurture"* (New York: Henry Holt, 2001).

Reid Barton's solution to problem five is described in "Top Young Problem Solvers Vie for Quiet Glory," by Dana Mackenzie, *Science* 293 (2001): 596–99.

CHAPTER 6. INTERLUDE
The game Twitch is distributed by Wizards of the Coast, P.O. Box 707, Renton, WA 98057.

CHAPTER 7. CREATIVITY
Among the many books written about creativity are *Creativity and Beyond: Cultures, Values, and Change,* by Robert Paul Weiner (Albany: State University of New York Press, 2000); *Encyclopedia of Creativity,* edited by Mark Runco and Steven Pritzker (New York: Academic Press, 1999); *Genius: The Natural History of Creativity,* by H. J. Eysenck (New York: Cambridge University Press, 1995); *Creativity: Beyond the Myth of Genius,* by Robert Weisberg (New York: W. H. Freeman, 1993); *Creating Minds: An Anatomy of Creativity Seen Through the Lives of Freud, Einstein, Picasso, Stravinsky, Eliot, Graham, and Gandhi,* by Howard Gardner (New York: Basic Books, 1993); *The Creative Mind: Myths and Mechanisms,* by Margaret Boden (New York: Basic Books, 1991); and *The Act of Creation,* by Arthur Koestler (New York: Macmillan, 1964).

Ravenna Helson and Richard Crutchfield describe the research on mathematicians and creativity at the Institute of Personality Assessment and Research in "Creative Types in Mathematics," *Journal of Personality* 38 (1970): 177–97, and "Mathematicians: The Creative Researcher and the Average Ph.D.," *Journal of Consulting and Clinical Psychology* 34 (1970): 250–57.

Dean Keith Simonton explores the parallels between biological evolution and creativity in *Origins of Genius: Darwinian Perspectives on Creativity* (New York: Oxford University Press, 1999). Another book by Simonton related to creativity is *Greatness: Who Makes History and Why* (New York: Guilford Press, 1994). As Simonton acknowledges, Donald Campbell proposed a Darwinian mechanism for creativity in "Blind Variation and Selective Retention in Creative Thought as in Other Knowledge Processes," *Psychological Review* 67 (1960): 380–400.

Simonton analyzes the productivity of classical composers in "Emergence and Realization of Genius: The Lives and Works of 120

Classical Composers," *Journal of Personality and Social Psychology* 61 (1991): 829–40.

The role of the subconscious in creativity is discussed by Jonathan Schooler and Joseph Melcher in "The Ineffability of Insight," pp. 97–134 in *The Creative Cognition Approach,* edited by Steven Smith, Thomas Ward, and Ronald Finke (Cambridge, Mass.: MIT Press, 1995).

Norman Maier presents his work on the string problem in "Reasoning in Humans, II: The Solution of a Problem and Its Appearance in Consciousness," *Journal of Comparative Psychology* 12 (1931): 181–94. See also his book *Problem Solving and Creativity in Individuals and Groups* (Belmont, Calif.: Brooks/Cole Publishing, 1970).

The Philip K. Dick quotation comes from the appendix of *Valis* (New York: Vintage, 1981). Sylvia Nasar writes about the relationship between John Nash's creativity and schizophrenia in the prologue to *A Beautiful Mind.*

Mihaly Csikszentmihalyi and Rick Robinson discuss the cultural component of creativity in "Culture, Time, and the Development of Talent," pp. 264–84 in *Conceptions of Giftedness,* edited by Robert Sternberg and Janet Davidson (New York: Cambridge University Press, 1986). Csikszentmihalyi further develops these ideas in "The Domain of Creativity," pp. 190–212 in *Theories of Creativity,* edited by Mark Runco and Robert Albert (Newbury Park, Calif.: Sage, 1990).

Asian-Americans: Achievement Beyond IQ, by James Flynn (Hillsdale, N.J.: Erlbaum, 1991), examines the controversy over Asian American success and intelligence. For a contrasting view, see "Oriental Americans: Their IQ, Educational Attainment, and Socio-Economic Status," by Richard Lynn, in *Personality and Individual Differences* (1993), pp. 237–42. See also "East-Asian Academic Success in the United States: Family, School, and Community Explanations," by Barbara Schneider, Joyce Hieshima, Sehahn Lee, and Stephen Plank, pp. 323–50 in *Cross-Cultural Roots of Minority Child Development,* edited by Patricia Greenfield and Rodney

Cocking (Hillsdale, N.J.: Erlbaum, 1994); and "Asian-American Educational Achievements: A Phenomenon in Search of an Explanation," by Stanley Sue and Sumie Okazaki, *American Psychologist* 45 (1990): 913–20.

Stacey Lee critically examines the impression of Asian American success in *Unraveling the "Model Minority" Stereotype: Listening to Asian American Youth* (New York: Teachers College Press, 1996).

Harold Stevenson and his colleagues analyze the cognitive differences between groups of children in Japan, China, and the United States in "Cognitive Performance and Academic Achievement of Japanese, Chinese, and American Children," *Child Development* 56 (1985): 718–34; and "Mathematics Achievement of Chinese, Japanese, and American Children: Ten Years Later," *Science* 259 (1993): 53–58.

For a comparison of immigrant children with youths whose families have been in the United States for several generations, see *From Generation to Generation: The Health and Well-Being of Children in Immigrant Families*, by the Committee on the Health and Adjustment of Immigrant Children and Families of the National Research Council and Institute of Medicine (Washington, D.C.: National Academy Press, 1998).

CHAPTER 8. BREADTH

The quotation from David Brooks comes from "One Nation, Slightly Divisible," *Atlantic Monthly* (December 2001), pp. 53–65.

Lewis Terman and his colleagues published a series of books over the course of half a century describing the characteristics and achievements of the Termites, including *Mental and Physical Traits of a Thousand Gifted Children* (Stanford, Calif.: Stanford University Press, 1925) and *The Gifted Child at Mid-Life* (Stanford, Calif.: Stanford University Press, 1959). A biography of Terman, a summary of his work, and profiles of several Termites appear in *Terman's Kids: The Groundbreaking Study of How the Gifted Grow Up*, by Joel Shurkin (Boston: Little Brown, 1992).

Howard Gardner discusses the origins of great achievement in "The Relationship Between Early Giftedness and Later Achievement," pp. 175–82 in *The Origins and Development of High Ability*, edited by Gregory Bock and Kate Ackrill (New York: Wiley, 1993). For an analysis of the formative experiences of seven prominent twentieth-century creators, see Gardner, *Creating Minds: An Anatomy of Creativity Seen Through the Lives of Freud, Einstein, Picasso, Stravinsky, Eliot, Graham, and Gandhi* (New York: Basic Books, 1993).

Some of the difficulties encountered by prodigies are described by Jeanne Bamberger in "Growing Up Prodigies: The Midlife Crisis," pp. 265–79 in *Developmental Approaches to Giftedness*, edited by David Feldman (San Francisco: Jossey-Bass, 1986). *Talented Teenagers: The Roots of Success and Failure*, by Mihaly Csikszentmihalyi, Kevin Rathunde, and Samuel Whalen, looks at a broader range of issues (New York: Cambridge University Press, 1993). For a notorious case of "early ripen, early rot," see "William James Sidis, the Broken Twig," by Kathleen Montour, *American Psychologist* 32 (1977): 265–79.

Robert Albert examines the career decisions made by high achievers in "Identity, Experiences, and Career Choice Among the Exceptionally Gifted and Eminent," pp. 13–34 in *Theories of Creativity*, edited by Mark Runco and Robert Albert (Newbury Park, Calif.: Sage, 1990). Ellen Winner describes the adult lives of gifted children in chapter 10 of *Gifted Children: Myths and Realities* (New York: Basic Books, 1996).

Ingrid Wickelgren describes Eric Lander's role in the Human Genome Project in *The Gene Masters: How a New Breed of Scientific Entrepreneurs Raced for the Biggest Prize in Biology* (New York: Times Books, 2002). Another useful summary is *Cracking the Genome: Inside the Race to Unlock Human DNA*, by Kevin Davies (New York: Free Press, 2001).

Most of James Campbell's published work on the math Olympians appears in a series of articles in the special issue "Cross-National Retrospective Studies of Mathematics Olympians," *Inter-*

national Journal of Educational Research 25 (1996). See especially "Early Identification of Mathematics Talent Has Long-Term Positive Consequences for Career Contributions," pp. 497–522. The Web site for the project is http://olympiadprojects.com/index.htm.

CHAPTER 9. A SENSE OF WONDER
Simon Singh tells the story of Fermat's last theorem in *Fermat's Enigma: The Epic Quest to Solve the World's Greatest Mathematical Problem* (New York: Walker, 1997). Singh also directed and coproduced the *Nova* television program "The Proof" that the Olympians watched. Another book describing the path to Wiles's solution is *Fermat's Last Theorem: Unlocking the Secret of an Ancient Mathematical Problem,* by Amir Aczel (New York: Four Walls Eight Windows, 1996). Accessible brief accounts of the proof appear in "Fermat's Last Stand," by Simon Singh and Kenneth Ribet, *Scientific American* (November 1997), pp. 68–73, and "Fermat's Last Theorem and Modern Arithmetic," by Kenneth Ribet and Brian Hayes, *American Scientist* 82 (1994): 144–56. For a more technical description of the proof, see "Galois Representations and Modular Forms," by Kenneth Ribet, *Bulletin of the American Mathematical Society* 32 (1995): 375–402. A personal chronology describing the process leading to the proof is *The Fermat Diary,* by Charles Mozzochi (Providence, R.I.: American Mathematical Society, 2000). The proof appears in two papers: "Modular Elliptic Curves and Fermat's Last Theorem," by Andrew Wiles, *Annals of Mathematics* 141 (1995): 443–551, and "Ring-Theoretic Properties of Certain Hecke Algebras," by Andrew Wiles and Richard Taylor, *Annals of Mathematics* 141 (1995): 553–72.

The book Wiles read as a child is *The Last Problem* by Eric Temple Bell (1961; reprint, Washington, D.C.: Mathematical Association of America, 1990). Manfred Schroeder describes some of the applications of number theory in "Number Theory and the Real World," *Mathematical Intelligencer* 7, no. 4 (1985): 18–26.

For an overview of the Langlands program, see "Fermat's Last Theorem's Cousin," by Dana Mackenzie, *Science* 287 (2000): 792–

93; and "New Heights for Number Theory," by Barry Cipra, *What's Happening in the Mathematical Sciences* 5 (2002): 2–11.

Keith Devlin describes the problems for which the Clay Mathematics Institute is offering $1 million rewards in *The Millennium Problems: The Seven Greatest Unsolved Mathematical Puzzles of Our Time* (New York: Basic Books, 2002). The problems are described more briefly in "Is That Your Final Equation?" by Charles Seife, *Science* 288 (2000): 1328–29; and "Think and Grow Rich," by Barry Cipra, *What's Happening in the Mathematical Sciences* 5 (2002): 76–87.

The Mathematical Experience, by Philip Davis and Reuben Hersh (Boston: Birkhäuser, 1981), contains many essays describing the beauty and utility of mathematics.

CHAPTER 10. TRIUMPH

The Olympiad programs in Russia and China are discussed in "The Olympiad Movement in Russia," by Boris Kukushkin, *International Journal of Educational Research* 25 (1996): 553–62; and "A Cross-Cultural Analysis of Similarities and Differences Among Math Olympiads in China, Taiwan, and the United States," by Alan Shoho, *International Journal of Educational Research* 25 (1996): 575–82. For more on the contrasting mathematical training of teachers in the United States and China, see *Knowing and Teaching Elementary Mathematics: Teachers' Understanding of Fundamental Mathematics in China and the United States*, by Liping Ma (Mahwah, N.J.: Erlbaum, 1999).

acknowledgments

This book could not have been written without the help and friendship of the six members of the 2001 U.S. Olympiad team — Reid Barton, Gabriel Carroll, Ian Le, Tiankai Liu, Oaz Nir, and David Shin — and of former team member Melanie Wood. I owe a special debt of gratitude to them all. They agreed to act as representatives for all those who have been on U.S. Olympiad teams, despite the time and effort they knew their cooperation would entail.

The coaches and administrators of the team were equally helpful. Titu Andreescu responded to endless phone calls and e-mail messages and allowed me to attend the summer training camp at Georgetown University. Zuming Feng and Zvezdelina Stankova gave me many insights into the competition, the competitors, and mathematics in general. Steve Dunbar at the American Mathematics Competitions and the University of Nebraska was a careful and encouraging guide and reviewer. Tina Straley and Donald Albers of the Mathematical Association of America, which sponsors the American Mathematics Competitions, provided invaluable assistance.

The International Mathematical Olympiad is a multimillion-dollar undertaking that could not occur without major financial contributions and the efforts of hundreds of volunteers. Walter Mientka, the executive director of IMO 2001, was responsible for the nearly impossible task of funding and staffing the event, yet he still had time to pick me up at the Lincoln airport and take me around to meet his colleagues. John Kenelly, the president of the group that oversaw the event, worked tirelessly to make the Olympiad a success. The International Mathematical Olympiad Jury allowed me to reproduce and discuss the six problems on the 2001 Olympiad. Among the major funders of the Olympiad were the Akamai Foundation, the Clay Mathematics Institute, Key Curriculum Press, the National Science Foundation, the National Security Agency, Renaissance Corporation, Texas Instruments, the U.S. Department of Education, and Wolfram Research.

Many people generously agreed to speak with me and to review parts of the manuscript. They include Pamela Barton, Joseph Bates, Camilla Persson Benbow, Steve Benson, George Berzsenyi, Hannah Burton, James Campbell, Beth Casey, Pam David, Peggy Drane, Sherry Eggers, Joan Ferrini-Mundy, Bob Fischer, Melvin George, Dick Gibbs, Kaining Gu, Brian Healy, James Hiebert, Arthur Jaffe, Alex Kasman, Kiran Kedlaya, Murray Klamkin, Eric Lander, Tri Le, Tom Leighton, Gordon Lessells, David Lubinski, Dana Mackenzie, Michael Mahoney, David Moore, Harold Reiter, Ken Ribet, Richard Rusczyk, Alex Saltman, Roger Shepard, Tatiana Shubin, Dean Keith Simonton, Tina Smith, Julian Stanley, James Stigler, Barbara Tompkins, Bob van Hoy, John Webb, Eric Wepsic, Steve Wildstrom, Susan Schwartz Wildstrom, Marcia Whatley, Ellen Winner, Paul Zeitz, and Wendy Ziner.

Once again my friend and literary agent Rafe Sagalyn made this book possible, both by immediately recognizing its potential and by continuing to encourage me over the many months of research and writing.

At the Mathematical Association of America, Lisa Kolbe helped me attend various Olympiad events. Bob and Sybil Rodman were gracious hosts in Palo Alto. Sidney and Adele Richman provided gracious accommodations for a revision of the manuscript.

At Houghton Mifflin, my editor, Laura van Dam, was instrumental in shaping the structure and contents of this book. The partnership I have with Laura is exactly the kind every writer could wish to have with an editor. Peg Anderson edited the manuscript with her usual grace and intelligence. Dan O'Connell enthusiastically organized publicity for the book. Erica Avery answered many important and not so important questions.

Finally, I want to thank Marci Jerina and the members of the math team that I help coach at Thomas Pyle Middle School in Bethesda, Maryland. Many of the observations made in this book derive from the Wednesday afternoons I spent with the team. Coincidentally, during the time when I was writing this book, the administration of the Montgomery County Public Schools decided to withdraw its support for the middle school math competitions that the county had sponsored for more than two decades. Fortunately, the competitions continued on a smaller scale, thanks to the volunteer efforts of a dedicated core of math teachers, including (in our district) Kathy Saftner, Eric Walstein, John Freed, Kathy Gaskill, and Marci Jerina. But school administrators who, in the name of equity, undermine the education of mathematically talented children are doing a disservice to their students, to their schools, and ultimately to themselves. It's a good thing kids are so resilient.

The love of my family — Lynn, Eric, and Sarah Olson — is reflected in every page of this book. I have dedicated the book to my mother, Diane Olson, who introduced me not only to mathematics but to the world of literature.